ANOMALOUS PHOTOCONDUCTIVITY

ANOMALOUS PHOTOCONDUCTIVITY

M. I. Korsunskii

Translated from Russian by Prof. E. Harnik
Racah Institute of Physics, Hebrew University of Jerusalem

A HALSTED PRESS BOOK

JOHN WILEY & SONS
New York · Toronto

ISRAEL PROGRAM FOR SCIENTIFIC TRANSLATIONS
Jerusalem · London

Sole distributors for the Western Hemisphere and Japan

HALSTED PRESS, a division of
JOHN WILEY & SONS, INC., NEW YORK

Library of Congress Cataloging in Publication Data

Korsunskiĭ, M. I.

    Anomalous photoconductivity.

    "A Halsted Press book."

    Translation of *Anomal'naia fotoprovodimost'*.

    1. Photoconductivity. 2. Semiconductors,
Effect of radiation on. I. Title.
QC612.P5K6613   537.5'4   73-12288
ISBN 0-470-50428-5

Distributors for the U. K., Europe, Africa and
the Middle East

JOHN WILEY & SONS, LTD., CHICHESTER

Distributed in the rest of the world by

KETER PUBLISHING HOUSE JERUSALEM LTD.
ISBN 0 7065 1369 X
IPST cat. no. 22086

This book is a translation from Russian of
ANOMAL'NAYA FOTOPROVODIMOST'
Izdatel'stvo "Nauka," Glavnaya Redaktsiya
Fiziko-Matematicheskoi Literatury
Moscow, 1972

Printed in Israel

PREFACE

This book investigates the nature and properties of the highly
curious and original phenomena of a n o m a l o u s   p h o t o c o n d u c -
t i v i t y   a n d   s p e c t r a l   m e m o r y   discovered in 1961 in specially
treated amorphous selenium layers.

Being one of a variety of photoconductivity phenomena in semiconductors, anomalous photoconductivity considerably widens the
horizon of these phenomena. It has led to the introduction of a new
type of centers in semiconductors, the s-centers, or storage centers,
whose functions markedly differ from those of recombination
centers and trapping levels.

A carrier captured by an s-center is stored in it apparently for
an indefinitely long time in the dark. It is shown experimentally
that the lifetime of a carrier in long-life traps in mercury-
activated amorphous selenium is longer (and, apparently, considerably longer) than one year. According to the subcolloidal model of
the s-center this lifetime is of the order of $10^{30}-10^{40}$ years.

In view of the existence of such centers, we have to extend and
refine the general theory of photoconductivity. It is evident that
the special features of the anomalous photoconductivity can be
utilized for the construction of a number of semiconductor devices
and detectors. All this encouraged us to systematize and generalize
the accumulated knowledge concerning anomalous photoconductivity
and s-centers.

The present book is intended for researchers engaged in the
study of photoelectrical phenomena in semiconductors, for post-
graduate students and solid-state electronics engineers interested
in photoconductivity phenomena, for students specializing in the
field of semiconductor physics and semiconductor devices.

M. I. Korsunskii

Contents

NOTATION

σ — conductivity of a sample

$\Delta\sigma_{ph}$ — photoresponse

σ_s — steady-state conductivity under illumination

σ_d — steady-state dark conductivity

$\Delta\sigma$ — instantaneous value of photoconductivity

σ_a — anomalous photoconductivity

$\sigma_a(\lambda)$ — steady-state valve of anomalous conductivity

σ_m — conductivity due to metastable carriers

$\sigma_m(\lambda)$ — steady-state value of conductivity due to metastable carriers

σ_n — nonamonalous (normal) component of photoconductivity

$\Delta\sigma_n$ — steady-state value of the nonanomalous (normal) component of photoconductivity (photoresponse of nonanomalous photoconductivity)

σ_n^d — dark value of the nonanomalous component of photoconductivity

σ_0 — conductivity due to equilibrium carriers

σ_i — conductivity of an intrinsic semiconductor

$\Delta\sigma^-$ — negative photoconductivity

$\Delta\sigma^+$ — positive photoconductivity

$\Delta\sigma_{ph}^+, \Delta\sigma_{ph}^-$ — photoresponse of positive and negative photoconductivity, respectively

σ_{max} — maximum value of conductivity

$\Delta\sigma_{max}$ — maximum value of photoconductivity

$\sigma^{(0)}$ — initial value of conductivity

$\tau_d^-, \tau_{d_i}^-$ — relaxation time of negative photoconductivity when the illumination is switched off

τ_0 — lifetime of equilibrium carriers

τ_q — lifetime of carriers in an s-center

$\tau(\lambda)$ — relaxation time of the anomalous component of photoconductivity

τ^+, τ_i^+ — relaxation time of positive photoconductivity

τ^-, τ_i^- — relaxation time of negative photoconductivity

$\tau_d^+, \tau_{d_i}^+$ — relaxation time of positive photoconductivity when the illumination is switched off

Q — concentration of s-centers responsible for anomalous photoconductivity

q — concentration of s-centers filled with excess carriers

Q^-, Q_i^- — concentration of s-centers responsible for the appearance of low-temperature negative photoconductivity in the semiconductor

q^-, q_i^- — concentration of above centers filled with excess carriers

Q^+, Q_i^+ — concentration of s-centers responsible for the appearance of positive photoconductivity in the semiconductor

q^+, q_i^+ — concentration of above centers filled with excess carriers

$q^-(\lambda), q(\lambda), q^+(\lambda)$ — steady-state value of concentration of s-centers filled with current carriers

γ — probability of trapping of excess carriers by s-centers responsible for anomalous photoconductivity

γ^-, γ_i^- — probability of trapping of excess carriers by s-centers responsible for negative photoconductivity

γ^+, γ_i^+ — probability of trapping of excess carriers by s-centers responsible for positive photoconductivity

$S, S(\lambda)$ — cross-section for the liberation of a carrier from an s-center by a light quantum

θ — probability of spontaneous ejection of a carrier from an s-center

n_0, p_0 — steady-state concentrations of equilibrium current carriers (electrons and holes, respectively)

m_i — concentration of metastable current carriers

$\Delta n, \Delta p$ — concentration of non-equilibrium (excess) current carriers

$\Delta n_{ss}, \Delta p_{ss}$ — steady-state values of the concentration of excess current carriers

n, p — concentration of electrons and holes, respectively

n_{ss}, p_{ss} — steady-state concentration of electrons and holes, respectively

β — quantum yield

β' — a quantity which has the meaning of quantum yield for the process of trapping excess carriers by s-centers

β_0, β_0' — relative values of respective yields

h — Planck's constant

$K, K(\lambda)$ — absorption coefficient of given radiation

$\varepsilon, h\nu$ — photon energy

E_g — width of forbidden band gap

E_i — excitation energy

ΔE — activation energy

E_F — Fermi level energy

E_r — energy of resonance level in an s-center

ΔE_r — width of resonance level

E_b — binding energy of a carrier in an s-center

D — transmission of a potential barrier

D_E — transmission of a potential carrier at energy level E

U — potential function

U_0 — height of potential barrier

a_0 — width of potential well

r_0 — radius of potential well

t — time

k — Boltzmann's constant

L — photon flux intensity

λ — wavelength

u_n, u_p — mobility of electrons and holes, respectively

u_m — mobility of metastable carriers

ρ — charge density

g — generation rate of equilibrium carriers

M — concentration of impurity centers other than long-life traps (ordinary impurity centers)

m — concentration of electron-filled centers of this type

m_0 — corresponding equilibrium concentration

γ_p — probability of hole capture by charged impurity center

m_{ss} — steady-state value of the concentration of charged centers

α — coefficient determining the rate of thermal generation of holes

α' — coefficient determining the rate of photogeneration of holes as a result of electron capture by impurity centers

μ — probability of absorption of a quantum by a carrier in an s-center

Z — color sensitivity of color resistance

R_f — film resistance

R_l — layer resistance

R_{opt} — optimal value of resistance

Chapter 1

MAIN FEATURES OF ANOMALOUS PHOTO-CONDUCTIVITY OF ACTIVATED AMORPHOUS SELENIUM

1. NORMAL AND ANOMALOUS PHOTOCONDUCTIVITY

It is well known that when a semiconductor is illuminated its electron conductivity changes in most cases. The magnitude of this conductivity change is expressed by the photoresponse $\Delta\sigma_{ph}$ which represents the difference between the steady-state value of the conductivity attained by an illuminated semiconductor σ_s and its dark value σ_d. An intrinsic characteristic of photoconductivity is the dependence of the photoresponse on the incident light intensity L. It is well known /1/ that the nature of this dependence differs in various semiconductors.

In addition to the simple dependences such as a) linear photo-conductivity, when the photoresponse varies as the first power of light intensity (linear or monomolecular recombination), b) cases when the photoresponse varies as the square root of light intensity (quadartic or bimolecular recombination), there are examples of exceedingly complex dependences of σ_s on L, mostly due to the existence in the sample of trapping centers, namely 1) superlinear photoresponse (when the photoresponse grows faster than the light intensity), 2) cases when the photoresponse depends very weakly on light intensity (for example, in manganese-doped germanium, when the intensity is changed by a factor of 100, the photoresponse changes only by a factor of two /2/).

There are cases when the conductivity of the semiconductor does not increase upon illumination but on the contrary decreases. The steady-state value of the conductivity in the presence of light is less than in the dark, and the more so the larger the illumination intensity. We shall subsequently call this type of photoconductivity n e g a t i v e in accordance with /3/, although some authors /4/ use the term a n o m a l o u s   p h o t o c o n d u c t i v i t y to refer exactly to this photoconductivity.

It is a common feature of all cases of photoconductivity considered up to now that the photoresponse is a function of the light intensity. Photoconductivity is therefore a photometric phenomenon, in which the illumination intensity is the leading factor. Each photoconductor has a certain characteristic spectral sensitivity (or spectral response), showing which spectral interval is the most effective. However, the spectral distribution of photoconductivity, which describes the wavelength dependence of the photoresponse $\Delta\sigma_{ph}$, is a comparison of the magnitudes of the photoresponse for different wavelengths but for the same light flux intensity L. The spectral response is thus a supplementary characteristic of a photoconductor, its main characteristic being the dependence of the photoresponse on light intensity. All the various forms of photoconductivity such that the photoresponse is a function of light intensity will subsequently be combined under the single term normal photoconductivity.

The term anomalous photoconductivity will be reserved for such cases in which the magnitude of the photoresponse does not depend on the light intensity but only on the spectral composition of the incident light. Such photoconductivity is not a photometric but a spectral phenomenon.

This form of photoconductivity was first observed in evaporated layers of amorphous selenium aged in an atmosphere of mercury vapor /5/. Samples which possess anomalous photoconductivity will be referred to as anomalously photoconductive semiconductors or anomalous photoconductors.

Anomalous photoconductivity has so far been observed only in amorphous selenium layers after a certain treatment. It was precisely in such samples that the features of anomalous photoconductivity were established and investigated. Therefore, in this book too, the properties of anomalous photoconductivity are examined on the basis of results obtained through the investigation of anomalously photoconductive amorphous selenium samples.

2. PREPARATION OF ANOMALOUSLY PHOTO-CONDUCTIVE SAMPLES OF AMORPHOUS SELENIUM

In this section we shall briefly describe the geometry of the sample, its preparation and methods of importing the anomalously photoconducting properties. These methods and the associated phenomena will be described in more detail in Chapter 5.

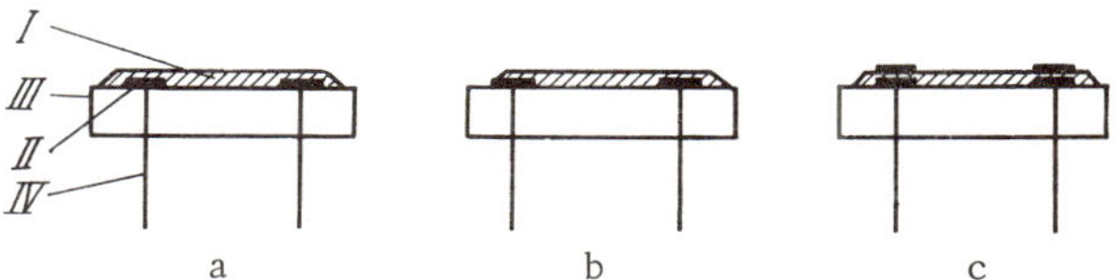

FIGURE 1. Schematic drawing of the samples used in the
investigation of anomalous photoconductivity:

a) both electrodes coated by selenium; b) one electrode
coated by selenium; c) electrodes fixed above and below
the selenium layer.

Amorphous selenium layers were obtained by evaporation in
vacuum using special quartz evaporation boats or tantalum boats.
The layers were deposited on glass or plexiglass plates. Gold
strips, vacuum-deposited on the substrate, served as electrodes.
The strips were blased to copper wires sealed or glued into the
substrate. Other kinds of electrodes, in particular silver or indium,
could be used instead of gold. Figure 1 shows schematically the
different samples of amorphous selenium layers used for the investigation of anomalous photoconductivity; here I denotes the
selenium layer, II the electrodes, III the substrate, and IV the wire
leads. The thickness of the selenium layer was of the order of $1\,\mu$m,
although some experiments were carried out using both thicker (up
to $5\,\mu$m) and thinner (up to $0.1\,\mu$m) layers. The electrodes were
3×6 mm parallel strips; the interelectrode separation varied
between 1 and 10 mm.

The samples most frequently used were selenium layers which
covered completely one of the gold electrodes and partly the second.
In some cases, the selenium layers completely covered both electrodes. Sometimes, a gold electrode was deposited on top of the
layer and connected with the electrode underneath.

The selenium layers prepared by this method did not show
anomalous photoconductivity. They acquired anomalously photoconductive properties, after further treatment (forming) /6/.

1. The layers were placed for some time in a closed vessel with
an open test tube containing mercury. There is an optimum time
for keeping the selenium layer in mercury vapor, and it depends on
the temperature. The treatment was kept under control by moni-
toring the conductivity of the sample. After the mercury-vapor
treatment, the resistance of the sample decreases by a factor of
10^5–10^6. We shall subsequently refer to the process of treating
selenium films with mercury vapor as film activation and the

films thus treated will be called m e r c u r y - a c t i v a t e d or simply a c t i v a t e d   f i l m s. The activation of samples by mercury is necessary but not sufficient for the appearance of anomalous photoconductivity. The samples must undergo further treatment.

2. The next stage is cooling. The sample is placed in a cryostat and is kept there throughout the experiment. The temperature of the selenium layer may be varied by means of a special oven from 100°K to room temperature. Anomalous photoconductivity was observed in the samples in the entire temperature interval below 200°K.

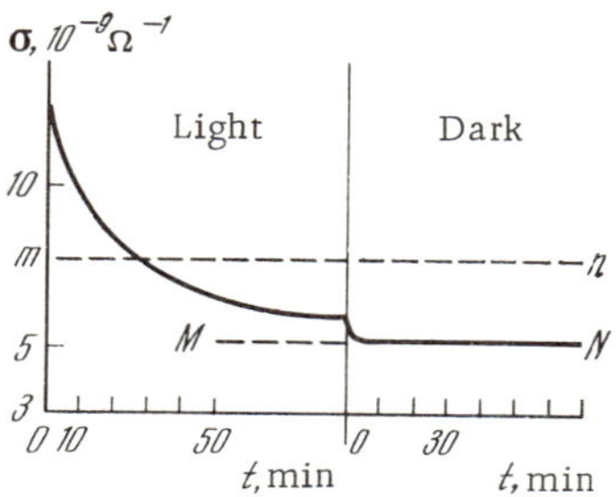

FIGURE 2. The change of conductivity during treatment with white light; mn is the dark conductivity of the sample before treatment, while MN is the dark conductivity after treatment

3. The last stage of the forming process is the illumination of the sample, kept below 200°K, by a high-intensity flux of white light. The illumination of selenium at 100—200°K leads to a residual change in the magnitude of its conductivity and photoconductivity. To obtain stable changes (i. e., to induce anomalous photoconductivity) it is necessary to expose the sample, kept at low temperature, to strong white light for sufficient length of time. During such a treatment, the conductivity of the sample changes continuously until it reaches some steady-state value (Figure 2). The dark con- ductivity of a treated sample does not differ much from its con- ductivity under illumination. Amorphous selenium films, activated by mercury cooled to sufficiently low temperatures and illuminated for a sufficient length of time, generally do show anomalous photo- conductivity. A more detailed discussion of the forming processes of an anomalous photoconductor will be given in Chapter 5.

3. THE PHOTORESPONSE OF ANOMALOUSLY PHOTOCONDUCTIVE SELENIUM LAYERS AS A FUNCTION OF LIGHT INTENSITY

It was mentioned in Section 1 that the photoresponse of an anomalous photoconductor does not depend on the light intensity.

Below, we shall present data illustrating this feature of anomalous photoconductivity. Note, however, that sometimes anomalous photoconductivity is mixed with a normal conductivity component. The method used to determine the dependence of the photoresponse $\Delta\sigma_{ph}$ on the light intensity L will be discussed in more detail later. Here we shall only present the results of the corresponding measurements /6/.

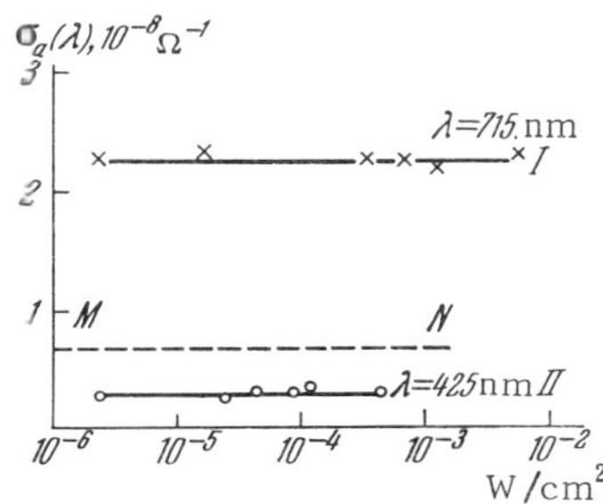

FIGURE 3. Anomalous photoconductivity of amorphous selenium vs. light intensity

In Figure 3, the ordinate is the magnitude of anomalous photoconductivity, and the abscissa is the light intensity on a logarithmic scale. The crosses denote the photoconductivity of a sample illuminated with red light ($\lambda = 715\,nm$) of the corresponding intensity, while the circles correspond to blue light ($\lambda = 425\,nm$). The crosses and the circles lie on two horizontal straight lines I and II. The dashed line MN represents the photoconductivity established in the sample following treatment by white light. The small fluctuations in the position of the experimental points are due to measurement errors.

Figure 3 proves that anomalous photoconductivity is independent of light intensity. The magnitude of the photoresponse remains the same, within the limits of experimental error (2%), as the intensity of red light is varied by almost four orders of magnitude. Note, however, that although the photoconductivity does not vary with light intensity when illuminated with either red ($\lambda = 715\,nm$) or blue ($\lambda = 425\,nm$) light, it does change (in the given sample, by almost an order of magnitude) when the wavelength of illumination is changed.

This shows that the constant value of the photoresponse over a wide range of light intensities cannot be trivially attributed to the impurity origin of photoconductivity assuming that over the whole range of intensities employed the impurity centers are saturated. If this were the case, the magnitude of the photoresponse would either be identical for illumination with red and blue light, or vary with the intensity for illumination with blue light ($\Delta\sigma_{blue} < \Delta\sigma_{red}$).

The fact that the magnitude of the photoconductivity increases substantially when the illumination is switched from blue to red shows that we are dealing not with saturation of impurity centers but with a different photoconductivity mechanism.

Our measurements of the photoresponse vs. light intensity were restricted to the range of $2 \cdot 10^{-6} - 6 \cdot 10^{-3}$ W/cm^2. This limitation is associated with difficulties that arise when the experiment is carried out in other intensity ranges. High light intensities cause notable heating of the sample and instability of its properties is observed; at low intensities the relaxation time increases considerably and very prolonged observations are necessary.

4. DARK CONDUCTIVITY OF ANOMALOUSLY PHOTOCONDUCTIVE SELENIUM

Anomalously photoconductive selenium largely retains its photoconductivity when the illumination is switched off. Figure 4 shows data obtained in one of the first experiments on anomalous photoconductivity /5/. The curves show the establishment of the steady-state value of photoconductivity in anomalously photoconductive amorphous selenium following its illumination with light of various wavelengths. The wavelength corresponding to each curve is indicated in the legend to the figure. In Figure 4 there are three regions separated by vertical lines. Region I shows the variation of photoconductivity under illumination, region II corresponds to dark conductivity, and region III shows a variation induced by changing the wavelength. The dashed line MN is the dark conductivity of the sample, established after forming treatment. It is clear that the photoconductivity induced by illumination is largely preserved when the light is switched off. This residual photoconductivity is preserved for a long time. Consequently, a sample of anomalously photoconductive selenium has an inherent spectral memory. The sample "remembers" for an indefinite length of time the spectral composition of the light to which it has been exposed.

Note that the small change in conductivity right after switching off the illumination (in some cases a drop in conductivity, in others a rise), which is observable on the oscillograms presented in Figure 4, is not characteristic of anomalous photoconductivity. The relative magnitude of such a change is different for different samples, for different physical conditions (intensity, temperature), and in some samples no such drop is observed. As will be shown later, such changes occur because the sample, in addition to anomalous

photoconductivity, also contains other components of photoconductivity. The true anomalous photoconductivity is precisely characterized by the value of the conductivity which is preserved after the
illumination is turned off.

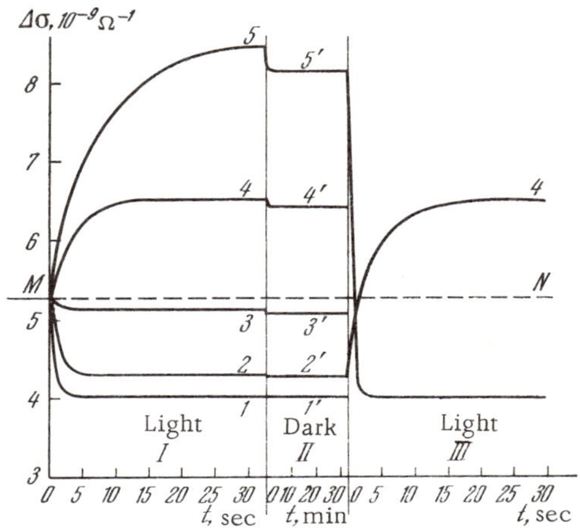

FIGURE 4. Photoconductivity relaxation curves:

1—5) wavelengths 425, 489, 530, 578, 715 nm, respectively.

How long can the conductivity induced in the sample by illumination be preserved in dark? An answer to this question is given by
the special long-duration conductivity observations carried out on a
series of anomalously photoconductive samples /7/. The sample
conductivity was recorded continuously for 100—120 hours (more
than four days). Results of observation on one of the samples
(No. 1VK) are given in Figure 5. The initial value of conductivity
$\sigma^{(0)}$ represents the dark conductivity established in the sample following illumination with blue light ($\lambda = 420\,\text{nm}$). The first part of
Figure 5 (curve I) illustrates the establishment of steady-state
photoconductivity under illumination with red light ($\lambda = 740\,\text{nm}$).
The second part (curve II) illustrates the establishment of dark
conductivity when this illumination is switched off. The time scale
of curves I and II is shown on the horizontal axis. The straight
line III is the level of steady-state conductivity ($\lambda = 420\,\text{nm}$). It is
seen that when the illumination is turned off, the conductivity keeps
on changing for a very long time. The sharp drop in conductivity
observable right after switching off the illumination is only a part
of a more prolonged fall which can be traced, within the accuracy
of measurements, over almost fifteen hours. Although the variation

of conductivity in the dark goes on for a long time, it clearly tends
to a definite limiting value, more than 10 times higher than the dark
value in the sample before its illumination with red light.   This
limiting conductivity which is attained after illumination is switched
off is in fact the steady-state value of the anomalous photoconductivity induced in the sample by the corresponding illumination.   It
is this value of conductivity which is independent of the light intensity incident on the sample, whereas the steady-state value of
the conductivity under illumination does depend on light intensity,
although very slightly.* It is possible to determine from the curve
in Figure 5 the lower limit of the relaxation time of anomalous
photoconductivity.   It will be shown in detail later on how this
estimate is obtained.   Here we shall only present the results: it
turns out that the relaxation time is over one and a half years.

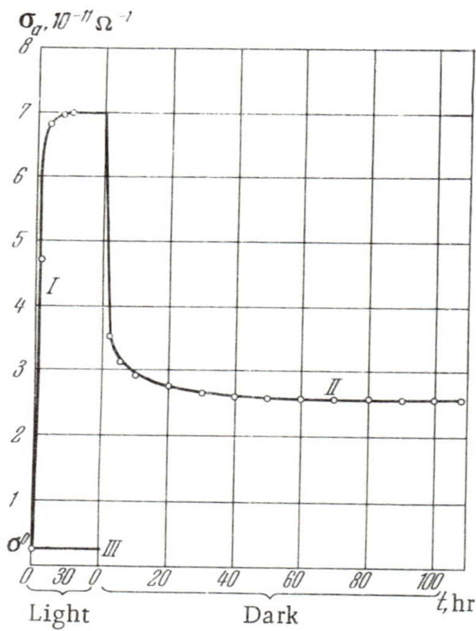

FIGURE 5. The relaxation of anomalous photoconductivity

It follows that the photoconductivity under illumination $\Delta\sigma$ can be
considered as consisting of two components:

$$\Delta\sigma = \sigma_a + \sigma_n,$$

where σ_a is the anomalous component of photoconductivity, while
σ_n is the component of some normal type of conductivity.   We shall

---

* It will be shown later that the difference between the conductivity established under illumination and the dark conductivity increases as the square root of light intensity.

subsequently denote by $\sigma_a(\lambda)$ and $\Delta\sigma_n$ respectively the steady-state components of photoconductivity under illumination.

When the illumination is switched off, the component σ_a remains unchanged whereas σ_n changes and approaches some "dark" value σ_n^d.

Since the anomalous photoconductivity induced by illumination depends on the wavelength and is preserved in the sample for a long, practically indefinite time after illumination is switched off, it is evident that in samples exhibiting anomalous photoconductivity we never encounter the true dark value of the sample conductivity.

The value of conductivity observable in a sample (in the dark) depends on the previous history of the sample and is therefore not characteristic of the sample. Therefore, the concept of dark con- ductivity has no meaning for an anomalously photoconductive sample. The dark value of the sample conductivity depends on the previous history of the sample, on the spectral composition of the prior illumination, and on the duration of illumination, and can vary by an order of magnitude. If the illumination of the sample is discontinued before the steady-state level is reached, the anomalous photoconductivity attained during the illumination stage is preserved in the dark.

Since the dark conductivity has no definite value in an anomalous photoconductor, the concept of photoresponse looses all meaning as applied to such semiconductors. The difference between the photo- conductivity established under illumination and its dark value depends on the previous history of the sample and on the spectral composition of the last illumination; consequently, it is not character- istic of the sample. We shall therefore not use "photoresponse" to characterize anomalous photoconductivity; the effect of illumina- tion will be characterized by the steady-state anomalous photo- conductivity $\sigma_a(\lambda)$ which, for a given sample, has a definite value depending only on the spectral composition of the light flux, but not on its intensity.

Samples exhibiting anomalous photoconductivity thus have a long- lived spectral memory. The sample remembers the magnitude of the light excitation, i.e., it retains the light-induced conductivity for a long time after the illumination is switched off.

5. SPECTRAL DISTRIBUTION OF ANOMALOUS PHOTOCONDUCTIVITY IN AMORPHOUS SELENIUM

It was mentioned in Sec. 1 that the steady-state value of anoma- lous photoconductivity under illumination $\sigma_a(\lambda)$, determined from the conductivity value observable when the light is switched off, depends on the spectral composition of the light and is a function of the incident wavelength.

Experimental results on the spectral distribution of anomalous photoconductivity are given in a number of publications /8—10/.

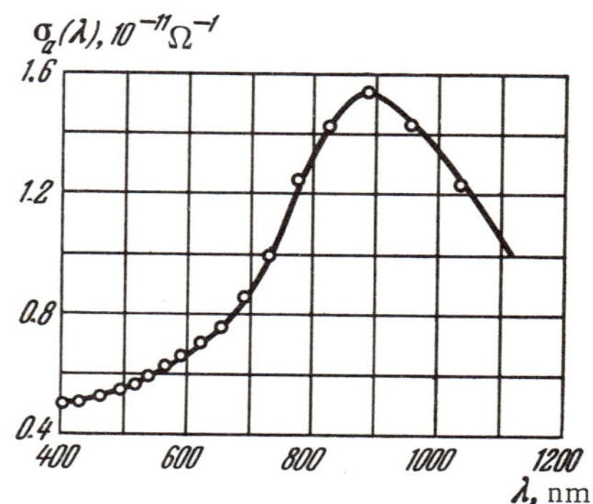

FIGURE 6. Spectral distribution of anomalous photoconductivity

The spectral distribution of anomalous photoconductivity is illustrated in Figure 6, where the abscissa represents the wavelength (in nm), and the ordinate — the steady-state value of anomalous photoconductivity. While remaining fairly constant in the region of the intrinsic absorption of amorphous selenium ($\lambda < 535$ nm), the photoresponse of the sample outside the absorption band increases sharply with the wavelength and reaches a maximum in the near-infrared region ($\lambda = 900$–$1,000$ nm).

Figure 7 shows the band structure of amorphous selenium, according to data published elsewhere /11/. The abscissa represents the density of electron states, while the ordinate gives the state energy. The band structure is such that in the ordered state the density of states is a parabolic function of energy, and the central section of the band bulges only in the glassy state. The band structure is plotted according to optical data, photoconductivity, and transport phenomena. The energy levels of the deep traps were determined from the magnitude of space-charge-limited currents.

According to this scheme, free electrons and holes are excited when light quanta with energy $\varepsilon = 2.53$ eV are absorbed. Absorption at $\varepsilon = 2.45$ eV is not associated with the formation of free charges. The Fermi level is distant 0.9 eV from the top of the valence band in selenium (for more details on the band structure of amorphous selenium, see Chapter 5, Sec. 1).

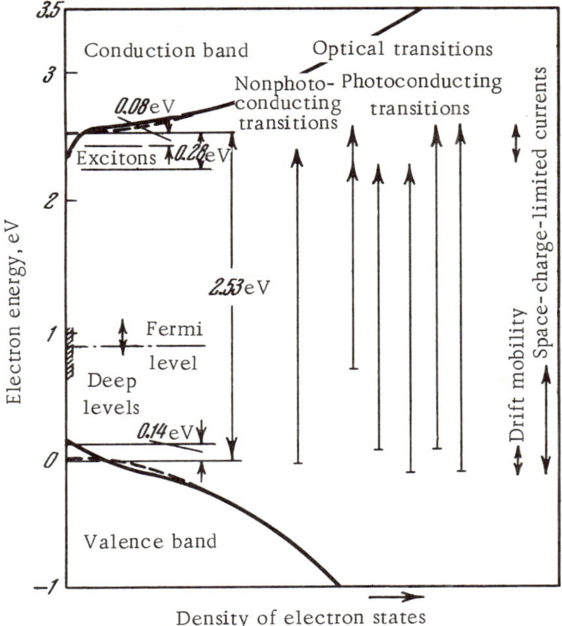

FIGURE 7. Electron states in amorphous selenium at room temperature, according to optical data, photoconductivity, and transport phenomena /11/. The solid curve corresponds to the glassy state, and the broken line to the ordered state.

Consequently, the maximum photoresponse of anomalous photoconductivity in selenium appears at a much longer wavelength than the absorption band edge of selenium. The energy of light quanta in the region of maximum photosensivity is less than half the width of the forbidden gap of amorphous selenium. The significance of this fact will be discussed in Chapter 4.

Let us mention here also some other features of the steady-state anomalous photoconductivity:

a) the steady-state anomalous photoconductivity established through illumination with a given wavelength (given spectral composition) does not depend on the value of the dark conductivity at the moment the light is switched on ($\sigma_a(\lambda)$ does not depend on the previous history of the sample);

b) the steady-state anomalous photoconductivity $\sigma_a(\lambda)$ (the spectral memory) responds fairly rapidly to any change in the spectral composition of the light flux (erasing the memory).

6. THE KINETICS OF CHANGES OF ANOMALOUS PHOTOCONDUCTIVITY FOLLOWING CHANGES IN THE SPECTRAL COMPOSITION OF ILLUMINATION

When the light wavelength is changed, the memory is erased, i.e., the conductivity of the sample changes. Some idea of the character of the resulting changes is given in Figure 4 (region III) above.

This figure shows, in particular, how the conductivity changes when the sample, which has been kept in the dark for a long time after having been exposed to light with $\lambda = 715\,\mathrm{nm}$, is illuminated with blue-violet light. The conductivity drops sharply and in a short time reaches a value corresponding to $\sigma_a(\lambda)$ for blue-violet light ($\lambda = 425\,\mathrm{nm}$).

Another feature of anomalous photoconductivity is that it can be either positive ($\sigma_s - \sigma_d > 0$) or negative ($\sigma_s - \sigma_d < 0$) in the same sample and even for illumination with the same light flux. Everything depends on the previous history of the sample. If the sample has previously been illuminated with light of a longer wavelength, illumination with light of the shorter wavelength will reduce the conductivity to a lower level (negative photoconductivity).

If the sample has previously been illuminated with shorter wavelength light, exposure to the longer wavelength leads to an increase in conductivity, which settles at a higher level (positive photoconductivity). Thus, for example, when a sample previously exposed to red light is illuminated with $\lambda = 600\,\mathrm{nm}$, the conductivity decreases, i.e., $\left(\dfrac{d\sigma}{dt}\right)_{t\to 0} < 0$. If, on the other hand, the same wavelength is used with a sample previously illuminated with blue light ($\lambda = 420\,\mathrm{nm}$), $\left(\dfrac{d\sigma}{dt}\right)_{t\to 0}$ will be positive (positive photoconductivity).

In the whole wavelength range $420-900\,\mathrm{nm}$, when the wavelength is changed from λ_i to λ_j, we have

$$\left(\frac{d\sigma}{dt}\right)_{t\to 0} > 0 \quad \text{for} \quad \lambda_j > \lambda_i,$$

and

$$\left(\frac{d\sigma}{dt}\right)_{t\to 0} < 0 \quad \text{for} \quad \lambda_j < \lambda_i.$$

This behavior of the anomalous photoconductivity is determined by the spectral memory the samples possess and by the spectral distribution of anomalous photoconductivity.

As mentioned above, the observed value of the photoconductivity of anomalously photoconductive amorphous selenium consists of two components representing the contributions of two photoconductivity mechanisms. In order to determine the kinetics of changes of anomalous photoconductivity, it is necessary to subtract the magnitude σ_n from the observed value of conductivity. This can be done in three ways.

1. Find the time dependence of $\Delta\sigma_n$ and decompose the curve $\Delta\sigma = f(t)$ into $\sigma_a = f_1(t)$ and $\sigma_n = f_2(t)$. The function $\sigma_n = f_2(t)$ can be found by analyzing the relaxation curves of the sample in the dark following its repeated illumination with a light flux of the same spectral composition.

2. Apply short-duration light pulses with long breaks between them. During the breaks, σ_n decreases, and since for small light doses the relative value of σ_n is in general small, the kinetics associated with such "pulsed" illumination represents the true kinetics of $\sigma_a = f_1(t)$.

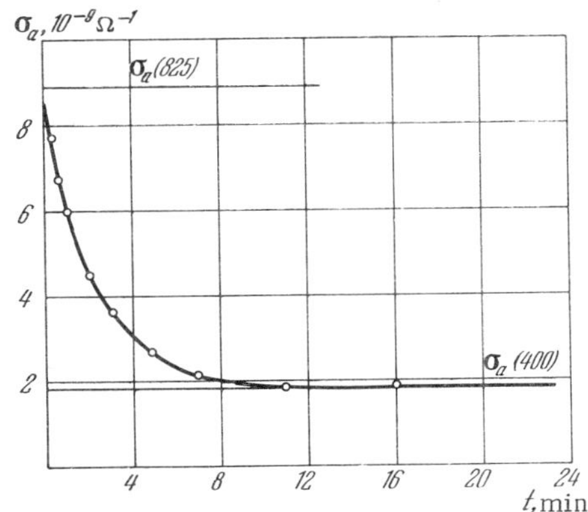

FIGURE 8. Relaxation of anomalous photoconductivity for pulsed illumination

3. Since the nonanomalous component $\Delta\sigma_n$ depends on light intensity, its contribution to conductivity becomes negligibly small when the intensity is appropriately reduced and the observed $\Delta\sigma = f(t)$ curve will represent the function $\sigma_a = f_1(t)$.

All these methods were employed for the determination of the relaxation time of anomalous photoconductivity and yielded consistent results. Figure 8 shows the variation of anomalous photoconductivity for pulsed illumination with $\lambda = 400\,\text{nm}$. This curve represents the relaxation of the anomalous component of photoconductivity, practically free from any nonanomalous component.

7. RELAXATION TIME OF ANOMALOUS PHOTOCONDUCTIVITY

The analysis of relaxation curves reveals that the variation of the conductivity of anomalously photoconductive selenium, induced by a change in the spectral composition of illumination, follows an exponential law.

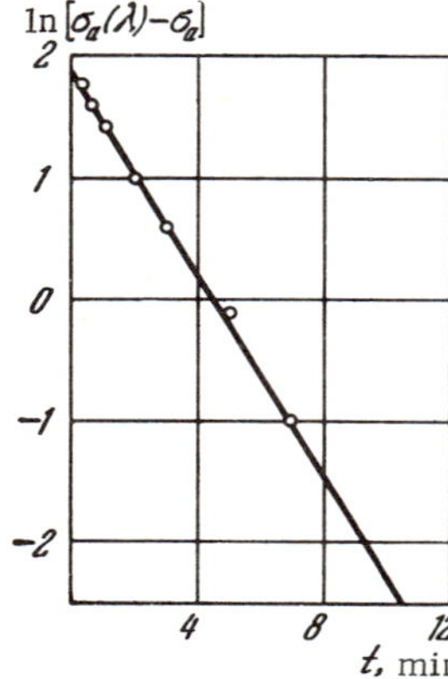

FIGURE 9. The determination of the relaxation time of photoconductivity

Figure 8 shows the relaxation curve of anomalous photoconductivity obtained by pulsed illumination of the sample. In Figure 9, the same relaxation curve is plotted as $\ln[\sigma_a(\lambda) - \sigma_a]$ vs. t. Its linear behavior confirms the above statement that the relaxation of anomalous photoconductivity is an exponential function of time:

$$\sigma_a(\lambda) - \sigma_a = [\sigma_a(\lambda) - \sigma^{(0)}]\, e^{-\frac{t}{\tau(\lambda)}},$$

$$\ln[\sigma_a(\lambda) - \sigma_a] = \ln[\sigma_a(\lambda) - \sigma^{(0)}] - \frac{t}{\tau(\lambda)}, \tag{1.1}$$

where $\sigma^{(0)}$ is the sample conductivity just before the illumination is switched on, $\tau(\lambda)$ is a magnitude which we shall call the r e l a x a t i o n t i m e o f a n o m a l o u s p h o t o c o n d u c t i v i t y f o r i l l u m i n a t i o n o f w a v e l e n g t h λ, or in short r e l a x a t i o n t i m e a t w a v e - l e n g t h λ; it characterizes the rate of establishment of the steady-state anomalous photoconductivity. The value of $\tau\,(\lambda)$ can be determined from the slope of the curve $\ln\,[\sigma_a\,(\lambda) - \sigma_a] = f\,(t)$, or from the relation

$$\tau\,(\lambda) = \frac{\sigma_a\,(\lambda) - \sigma^{(0)}}{\left(\dfrac{d\sigma_a}{dt}\right)_{t\to 0}} \, . \qquad\qquad (1.2)$$

Note that $\tau\,(\lambda)$ is always positive for $\left(\dfrac{d\sigma_a}{dt}\right)_{t\to 0} > 0$, if $\sigma_a\,(\lambda) > \sigma^{(0)}$, and $\left(\dfrac{d\sigma_a}{dt}\right)_{t\to 0} < 0$ if $\sigma_a\,(\lambda) < \sigma^{(0)}$.

We shall now consider some properties characteristic of the magnitude $\tau\,(\lambda)$.

a) $\tau\,(\lambda)$ does not depend on the previous history of the sample and it has a constant value for given experimental conditions, i. e., for given spectral composition and illumination intensity. This is illustrated in Figure 10, taken from /6/. The figure shows relaxation curves for illumination at $\lambda = 580\,\text{nm}$. Curve I corresponds to the relaxation of anomalous photoconductivity at this wavelength from a dark photoconductivity 1', established following illumination with blue light ($\lambda = 420\,\text{nm}$). Curve II represents relaxation at this wavelength from a dark value 4' reached following illumination of the sample with red light ($\lambda = 640\,\text{nm}$). The straight line III is the level of conductivity attained following illumination at $\lambda = 580\,\text{nm}$. Although in one case the photoconductivity is positive while in the other case it is negative, and the absolute value of $\sigma_a\,(\lambda) - \sigma^{(0)}$ is different in the two cases, the magnitude of $\tau\,(\lambda)$ for the two relaxation curves is the same, being approximately 6 sec. $\tau\,(\lambda)$ is thus indeed independent of previous history, and is determined only by the wavelength of the incident radiation and its intensity.

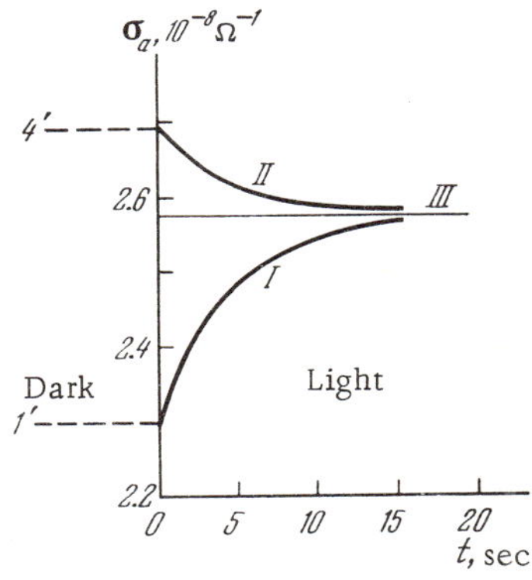

FIGURE 10. Independence of $\tau\,(\lambda)$ on the previous history of the sample

b) The relaxation time $\tau(\lambda)$ depends on the light intensity. Table 1 presents numerical values of $\tau(\lambda)$ from /6/, for relaxation at $\lambda = 420$ nm and $\lambda = 640$ nm for various light intensities L.

TABLE 1

λ, nm	L, W/cm^2	$\tau(\lambda)$, sec	$\tau(\lambda)\cdot L$, Wsec/cm^2
640	$2.4\cdot10^{-6}$	1100	$2.7\cdot10^{-3}$
	$1.7\cdot10^{-5}$	130	$2.3\cdot10^{-3}$
	$7.5\cdot10^{-5}$	34	$2.5\cdot10^{-3}$
	$1.0\cdot10^{-4}$	23	$2.3\cdot10^{-3}$
	$1.5\cdot10^{-4}$	16	$2.4\cdot10^{-3}$
	$2.0\cdot10^{-4}$	13	$2.6\cdot10^{-3}$
	$4.2\cdot10^{-4}$	6	$2.5\cdot10^{-3}$
420	$1.0\cdot10^{-4}$	0.29	$2.9\cdot10^{-5}$
	$1.7\cdot10^{-5}$	1.8	$3.1\cdot10^{-5}$
	$8.4\cdot10^{-5}$	3.2	$2.7\cdot10^{-5}$
	$2.5\cdot10^{-6}$	11.8	$2.7\cdot10^{-5}$

Over the entire intensity range (more than two orders of magnitude) the product $\tau(\lambda)\cdot L$ (Table 1) remains constant, within the limits of experimental error. Consequently, the relaxation time $\tau(\lambda)$ is a hyperbolic function of the intensity

$$\tau(\lambda) = \frac{\tau^{(0)}(\lambda)}{L}, \tag{1.3}$$

where $\tau^{(0)}(\lambda)$ is a magnitude characteristic of the spectral composition of radiation, and is a function of the wavelength only. This function will be considered in Chapter 3.

c) It follows from (1.2) and (1.3) that the instantaneous value of anomalous photoconductivity σ_a is determined by the light dose that the sample has received in the illumination process. Indeed, substituting (1.3) in (1.1) we obtain

$$\sigma_a(\lambda) - \sigma_a = [\sigma_a(\lambda) - \sigma^{(0)}]\, e^{-\frac{t\cdot L}{\tau^{(0)}(\lambda)}}. \tag{1.4}$$

Thus, the anomalous photoconductivity σ_a attained by the time t is determined by the product $L\cdot t$, i.e., by the radiation dose.

The value of σ_a depends not only on the radiation dose but also on the dark conductivity level from which the change begins. However, for each type of relaxation ($\lambda_i \rightarrow \lambda_j$), the instantaneous value of anomalous photoconductivity is entirely determined by the radiation dose. Consequently, the difference $\sigma_a(\lambda) - \sigma_a$ in this case can serve as a measure of the radiation dose.

d) The fact that the relaxation time is inversely proportional to light intensity leads to experimental difficulties in the determination of $\sigma_a(\lambda)$ with low light fluxes.

It is evident from the data in Table 1 that already at a red-light intensity $L = 2 \cdot 10^{-6}$ W/cm^2, the relaxation time $\tau(\lambda)$ is $1,100$ sec, i. e., some two hours are required for $\sigma_a(\lambda)$ to reach its steady-state value under illumination. It is therefore difficult to extend the function $\sigma_a(\lambda) = f(t)$ far toward low intensities.

e) Let us compare the relaxation time from λ_j to λ_i (i. e., relaxation from the photoconductivity level attained with illumination at wavelength λ_j to the pre-conditioning level at λ_i) with the opposite case of relaxation from λ_i to λ_j. The absolute value of the photoconductivity increment is the same in the two cases; the relaxation time at $\lambda_j, \tau(\lambda_i)$, and that at $\lambda_i, \tau(\lambda_j)$, are different, however.

Therefore, the magnitudes

$$\left(\frac{d\sigma}{dt}\right)_{t\to0}^{\lambda_i \to \lambda_j} = \frac{\sigma_a(\lambda_j) - \sigma_a(\lambda_i)}{\tau(\lambda_j)}$$

and

$$\left(\frac{d\sigma}{dt}\right)_{t\to0}^{\lambda_j \to \lambda_i} = \frac{\sigma_a(\lambda_i) - \sigma_a(\lambda_j)}{\tau(\lambda_i)} = -\frac{\sigma_a(\lambda_j) - \sigma_a(\lambda_i)}{\tau(\lambda_i)}$$

have different signs and, moreover, differ in their absolute values.

Let us summarize and enumerate the main properties of anomalous photoconductivity.

1. $\sigma_a(\lambda)$ does not depend on light intensity L.

2. $\sigma_a(\lambda)$ (while independent of L) is a function of the incident wavelength.

3. In an anomalous photoconductor, the true dark conductivity cannot be determined. This concept is meaningless here.

4. There is no meaning to the concept of photoresponse either. For a given sample and a given light flux, the photoresponse is indeterminate. It depends on previous history of the sample. Anomalous photoconductivity is uniquely characterized by the steady-state value of conductivity which is established in the sample and can be measured by the steady-state conductivity observed after switching off the illumination. $\sigma_a(\lambda)$ does not depend on previous history or on the value of conductivity prior to illumination.

5. The spectral distribution of $\sigma_a(\lambda)$ differs significantly from the photoresponse of nonactivated selenium. Crossing the absorption edge of amorphous selenium from short to long wavelength, the

anomalous photoconductivity does not decrease: in fact it increases reaching a maximum at $\lambda = 900\,\text{nm}$. On the other hand, the transmittance curve of light in anomalously photoconductive selenium films is the same as in normally photoconductive films.

6. The spectral distribution of anomalous photoconductivity described in the previous paragraph (item 5) is responsible for the fact that in the same sample, depending on previous history, the anomalous photoconductivity is positive $\left(\left(\frac{d\sigma}{dt}\right)_{t\to 0} > 0\right)$ for one set of illumination conditions (the wavelength of incident light is longer than that of prior illumination) and negative $\left(\left(\frac{d\sigma}{dt}\right)_{t\to 0} < 0\right)$ for other conditions (the wavelength of prior illumination is longer than that of the incident light). Thus, $\left(\frac{d\sigma}{dt}\right)_{t\to 0}$ will have different signs for the spectral composition of radiation, changing in the opposite directions ($\lambda_i \to \lambda_j$ and $\lambda_j \to \lambda_i$). Moreover, the absolute values of $\left(\frac{d\sigma}{dt}\right)_{t\to 0}$ will differ for these two courses of light flux varieties.

7. Anomalously photoconductive semiconductors possess a "color memory." The anomalous photoconductivity established under illumination of the sample is preserved practically an infinitely long time after the illumination is switched off. The quasi-dark conductivity observed in this way is a function of the wavelength to which the sample had been exposed.

8. When the wavelength of the incident light is changed, the "memory" is relatively quickly erased and as a result a new steady-state value $\sigma_a(\lambda)$ is established.

9. When the illumination wavelength is changed from λ_i to λ_j, the anomalous photoconductivity changes according to an exponential law

$$\sigma_a = \sigma_a(\lambda_j) - [\sigma_a(\lambda_j) - \sigma_a(\lambda_i)]\, e^{-\frac{t}{\tau(\lambda_j)}}.$$

10. The relaxation time $\tau(\lambda)$ does not depend on the magnitude of the quasi-dark conductivity (i.e., it does not depend on previous history) and is determined only by current illumination.

11. For constant L, $\tau(\lambda)$ changes abruptly when the wavelength is changed.

12. For a given λ, $\tau(\lambda)$ is a function of the light intensity. The product $\tau(\lambda)\cdot L$ is constant. The level of conductivity established in the sample is determined by the light dose.

Chapter 2

PHENOMENOLOGICAL THEORY OF
ANOMALOUS PHOTOCONDUCTIVITY

1. THE HYPOTHESIS OF LONG-LIFE TRAPS (STORAGE CENTERS)

According to Rose /1, 13, 14/, the photoconductivity is determined by local centers which exist in the semiconductor. These centers can be divided into recombination centers and trapping centers, according to the role they play in photoconductivity. Carriers trapped on trapping levels do not recombine with free carriers of the opposite sign: the probability of such a recombination is much lower than the probability of transition of the carrier to the appropriate band due to its thermal energy. Carrier recombination takes place mainly via recombination centers. The photoconductivity of a semiconductor is determined by the concentration of recombination and trapping centers and their cross-section for interaction with carriers. However, the existence of these centers alone does not explain the phenomenon of anomalous photoconductivity. The properties of anomalously photoconductive semiconductors described in Chapter 1 can be explained if it is assumed that the usual thermal diffusion recombination processes are absent in these semiconductors, and that carriers trapped in trapping centers are not freed through recombination with free carriers of the opposite sign, nor through thermal excitation into the appropriate band, but by quanta of the incident light /15, 16/.

It is easy to show what basic properties such centers must possess so that their presence in the semiconductor should lead to anomalous photoconductivity.

1. Carriers trapped in such centers cannot jump into the appropriate band by thermal motion alone. The probability of such an excitation is so small that it can be neglected. Consequently, such centers are not trapping centers.

2. Carriers trapped in such centers cannot recombine with free carriers of opposite sign. The cross-section for such recombination is so small (in anomalously photoconductive selenium, it is

less than 10^{-32} cm^2) that it can be neglected. Consequently, the centers under consideration are not recombination centers.

Such local centers in which there is no carrier recombination and from which no thermal excitation into the appropiate band takes place will be called storage centers (s-centers for short) or long-life traps. The last term was used in a number of publications in which the properties of anomalous photoconductivity were first considered.

Long-life traps (s-centers) must therefore possess the following basic properties.

1. A carrier hitting a long-life trap is stored in it for an indefinite time in the dark.

2. Equilibrium carriers are not trapped by s-centers; the probability of such capture is so small that it can be neglected.

Only combination of the two properties can explain the fact that the conductivity induced in the semiconductor by illumination remains unchanged for an indefinitely long time after the illumination is switched off ("spectral memory").

3. A carrier trapped by an s-center can be liberated by absorbing a light quantum.

Only thus is it possible to "erase the memory," i. e., to change the conductivity by changing the spectral composition of illumination.

These are the three necessary properties which local centers responsible for the appearance of anomalous photoconductivity must have. It will be shown later that these properties are sufficient to explain the features of anomalous photoconductivity enumerated in Chapter 1.

The term long-life trap (or s-center) stresses that a carrier is stored in the trap for an indefinitely long time in the dark (for more than one and a half years in anomalously photoconductive selenium).

It is clear that such a prolonged storage of a carrier is due to the peculiar structure and properties of the long-life trap. These properties will be discussed in Chapter 3.

There are apparently long-life traps capable of storing electrons, and also traps capable of storing holes. In principle, anomalous photoconductivity may be of either electron or hole origin. In anomalously photoconductive selenium, hole-type (p-type) conductivity is apparently observed. Therefore, in future, when considering the properties of semiconductors containing s-centers, we shall assume that the s-centers capture electrons (excess carriers). All that follows remains valid for electron-type (n-type) anomalous photoconductivity; it is only necessary to assume that the long-life traps capture holes, not electrons.

To clarify the role of s-centers in photoconductivity, we shall consider a semiconductor which contains only one type of impurity centers, namely s-centers. We shall establish what photoconductivity such a semiconductor should possess and find out whether it can behave like an "anomalous photoconductor."

2. SPECIFIC FEATURES OF THE PHOTO-CONDUCTIVITY OF A SEMICONDUCTOR WITH S-CENTERS

Let us assume, following /16/, that the semiconductor contains only one type of impurity centers, namely s-centers. To fix ideas, we shall assume that the long-life traps capture electrons. When such a semiconductor is illuminated, the occupation of long-life traps begins to change. Free electrons generated in the semiconductor as a result of absorption of light quanta are captured by empty traps, not occupied by electrons (filling the traps). On the other hand, absorption of light quanta knocks out electrons from originally filled traps (emptying the traps). The liberated electrons recombine relatively fast with free holes, correspondingly reducing the conductivity of the semiconductor.

The electronic transitions scheme corresponding to this model is given in Figure 11, where L is the number of quanta incident each second on 1 cm^2 of the semiconductor surface, K is the absorption coefficient of selenium for the given monochromatic radiation, τ_0 is the lifetime of free carriers in the semiconductor (in amorphous selenium, according to Moss /12/, τ_0 is of the order of 10^{-7} sec), Q is the concentration of s-centers, q is the concentration of s-centers containing a trapped electron (filled long-life traps), γ is the probability for the capture by an s-center of a free electron from the conduction band, S is the partial cross-section for liberation of an electron from an s-center by a light quantum and its subsequent recombination with a free hole, n is the concentration of free electrons in the conduction band, β is the quantum yield.

Note that the magnitudes K, S and, possibly, β and γ are functions of the wavelength of incident light.

The electron transitions scheme of Figure 11 differs from the usual schemes with a single class of impurity centers in that it allows no direct recombination of a free hole with an electron in a trap (the recombination cross-section is assumed to be zero), and no thermal excitation of electrons from long-life traps into the conduction band.

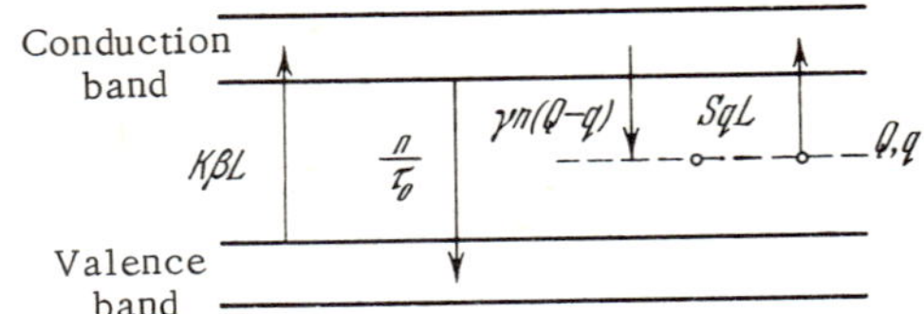

FIGURE 11. Electron transitions scheme in a
semiconductor containing long-life traps

The kinetic equations in this case have the following form:

$$\frac{dn}{dt} = K\beta L - \frac{n}{\tau_0} - \gamma n\,(Q - q) + SqL, \tag{2.1}$$

$$\frac{dq}{dt} = \gamma n\,(Q - q) - SqL \tag{2.2}$$

and

$$p = n + q \tag{2.3}$$

(p is the concentration of free holes in the valence band). In equilibrium, we have

$$\gamma n_{ss}\,(Q - q_{ss}) = S\,(p_{ss} - n_{ss})\,L, \tag{2.4}$$

$$n_{ss} = K\beta L\tau_0, \tag{2.5}$$

where n_{ss}, p_{ss}, q_{ss} are respectively the steady-state values of the
concentration of free electrons, holes and electrons in traps. In
equations (2.1)–(2.5), the probability for the penetration of an equilibrium electron into an s-center γn_0 is taken to be zero. It follows
from (2.4) and (2.5) that

$$p_{ss} = \frac{K\beta\tau_0\gamma}{S + K\beta\tau_0\gamma}\,Q + K\beta\tau_0 L. \tag{2.6}$$

If moreover

$$SL + \gamma K\beta\tau_0 L \ll \gamma Q, \tag{2.7}$$

the concentration of holes in the valence band becomes independent
of the quantum flux intensity L.

When condition (2.7) is fulfilled, we have

$$p_{ss} = \frac{K\beta\tau_0\gamma}{S + K\beta\tau_0\gamma}\,Q. \tag{2.8}$$

Although the hole concentration in the valence band is independent of light intensity when condition (2.7) is fulfilled, the electron concentration in the conduction band is a function of L (see (2.5)). In this case, the ratio of n_{ss} to p_{ss} is given by

$$\frac{n_{ss}}{p_{ss}} = \frac{SL + \gamma K\beta\tau_0 L}{\gamma Q} \ll 1. \tag{2.9}$$

Thus, when condition (2.7) is fulfilled, the concentration of free electrons in the conduction band is small compared with the concentration of holes in the valence band, and since the electron mobility in amorphous selenium /17/ is considerably lower than the hole mobility, the electron contribution to conductivity can be neglected. Consequently, the photoconductivity

$$\sigma_s = eu_p p_{ss} = eu_p \frac{K\beta\gamma\tau_0}{S + K\beta\gamma\tau_0} Q \tag{2.10}$$

is independent of light intensity. In (2.10) u_p is the hole mobility.

Thus, a semiconductor containing only one type of impurity centers, i.e., long-life traps or s-centers, from which electrons are knocked out by light quanta, possesses one of the basic properties of anomalous photoconductivity under inequality (2.7), namely σ_s is independent of L (property 1 in the account of properties of anomalous photoconductivity at the end of Chapter 1).

Note that the independence of σ_s on L can be obtained in two ways. Indeed, since $n_{ss} \ll p_{ss}$, we have

$$p_{ss} \simeq q_{ss} = \frac{K\beta\tau_0\gamma}{S + K\beta\tau_0\gamma} Q. \tag{2.11}$$

If moreover

$$K\beta\tau_0\gamma \gg S, \tag{2.12}$$

then

$$p_{ss} \simeq q_{ss} = Q, \tag{2.13}$$

i.e., when condition (2.12) is fulfilled, it appears that the steady-state photoconductivity is independent of light intensity for a trivial reason, namely because all the traps are filled. Since (2.13) does not depend on K, β, S, γ and consequently on λ, condition (2.13) cannot hold true for anomalous photoconductivity. Anomalous photoconductivity can appear in a semiconductor when the opposite condition obtains

$$K\beta\tau_0\gamma \ll S, \tag{2.14}$$

which indicated that the probability for the penetration of a free electron into a trap is less than the probability for an electron to be knocked out by a light quantum from a trap.

In this case

$$q_{ss} \simeq p_{ss} \simeq \frac{K\beta\tau_0\gamma}{S}Q \ll Q. \tag{2.15}$$

Inequality (2.15) signifies that in steady-state conditions, when (2.14) holds true, only part of the long-life traps are filled. The lower the probability for the penetration of a free electron into a long-life trap compared with the probability for its ejection from a long-life trap by a light quantum, the smaller the proportion of filled traps.

The fact that steady-state conductivity under illumination σ_s is independent of light intensity is thus the result of a dynamic equilibrium between the number of electrons trapped in long-life traps following the absorption of light quanta and the number of electrons knocked out from long-life traps back into the conduction band, due to absorption of light quanta by filled traps. The steady-state conductivity under illumination in this case is given by

$$\sigma_s = eu_p p_{ss} = eu_p \frac{K\beta\tau_0\gamma}{S}Q. \tag{2.16}$$

Since K, S and, possibly, β and γ are functions of the wavelength, the steady-state conductivity under illumination for a semiconductor with long-life traps, while independent of light intensity, is a function of the wavelength. This is another of the properties of anomalous photoconductivity (property 2 in the account at the end of Chapter 1).

Relation (2.16) leads to the conclusion that the steady-state conductivity under illumination in such a semiconductor will depend not only on the wavelength but also on the properties of the sample, in particular on the concentration Q of long-life traps. Although the spectral distribution curves of anomalous photoconductivity are similar in all the samples investigated, the absolute values of their conductivity may differ appreciably, especially as a result of differences in Q (which is indeed so).

The lifetime of free carriers in amorphous selenium /12/ is small, of the order of 10^{-7} sec. With such a short lifetime, free electrons (in the conduction band) recombine with holes shortly after the illumination is switched off and the concentration n vanishes.* Since there are no free electrons in the dark, the trapping of electrons by s-centers ceases. As $n_{ss} \ll p_{ss}$, the recombination of free holes with free electrons hardly changes the magnitude of free hole concentration and, consequently, does not change the conductivity either.

Therefore, after the illumination is switched off, and free electrons recombine with free holes, the flow of electrons into long-life traps and the ejection of electrons from filled traps both cease. Since no holes are produced or destroyed after the illumination is switched off, the conductivity of the semiconductor retains the same value as that reached under illumination. In other words, a semiconductor containing s-centers must possess spectral memory (property 7 in the account of properties of anomalous photoconductivity), i. e., it behaves as an anomalous photoconductor.

The conditions for the appearance of anomalous photoconductivity determined by relations (2.7) and (2.14) can be combined into the single condition (2.17). Note that inequality (2.14) can be relaxed ($K\beta\tau_0\gamma < S$, and not $K\beta\tau_0\gamma \ll S$). In this case, however, the wavelength dependence of the steady-state conductivity will be less pronounced.

The conditions

$$\gamma n_{ss} = K\beta\tau_0\gamma L < SL \ll \gamma Q \qquad (2.17)$$

indicate that in a semiconductor containing s-centers, anomalous photoconductivity can appear only when:

1) the concentration of excess carriers (electrons) which can be trapped by these centers is small compared with the concentration of carriers of the opposite sign (holes);

2) the probability of a free carrier being trapped is small compared with the probability of a carrier being knocked out from a trap by a light quantum.

Condition (2.17) in a semiconductor with long-life traps also leads to the appearance of spectral memory (i. e., σ_s independent of L). The shape of the spectral distribution curve does not contradict our model. However, since the nature of long-life traps has not yet been established, the dependence of S and γ on λ cannot be predicted. On the basis of the fact that σ_s increases with the wavelength (rather than decreases), it can be concluded that in the given spectral interval S varies with wavelength faster than the product $K\beta\gamma$. As a result the steady-state photoconductivity in blue light, i. e., in the region of the intrinsic absorption of selenium, is

---

* Apparently, in activated amorphous selenium films, γn_0 is small at temperatures at which anomalous photoconductivity exists because n_0 is small.

considerably less than that in the long-wave part of the spectrum, which is relatively weakly absorbed by selenium.

Properties 3 and 4 of anomalous photoconductivity enumerated in Chapter 1 also follow from our model. Indeed, since the concentration of electrons in long-life traps remains constant for an indefinitely long time, after the light is switched off, the experimentally observed dark conductivity of a semiconductor with long-life traps depends on previous history of the sample, i. e.. on wavelength and duration of the previous illumination. At the same time, the magnitude of conductivity established under illumination, determinable by (2.10), does not depend on previous history of the sample, nor on the particular value of the quasi-dark conductivity when the illumination was switched off. Consequently, in a semiconductor with long-life traps, photoresponse and dark conductivity have no meaning.

The dark conductivity in anomalous photoconductors is variable and dependent on previous history because the free carriers created in these semiconductors by illumination are preserved in the dark for an indefinitely long time. These are not equilibrium carriers (although they exist for an indefinitely long time). Yet they differ from the usual excess carriers (whose main characteristic is that they disappear after the illumination is switched off) in the absence of recombinations and their long lifetime in the dark. Free carriers which are produced as a result of the capture of excess carriers by s-centers will be called m e t a s t a b l e   c a r r i e r s.

Anomalous photoconductivity is associated with the appearance of such metastable carriers and is essentially metastable photoconductivity. Metastable photoconductivity excited by illumination is preserved after illumination is switched off.

In order to clarify whether the remaining properties (7—12) of anomalous photoconductivity are also inherent to a semiconductor with long-life traps, we shall examine the kinetics of the establishment of steady-state conductivity in a sample when the spectral composition of the incident light is changed.

3. KINETICS OF THE ESTABLISHMENT OF STEADY-STATE CONDUCTIVITY IN SEMICONDUCTORS WITH LONG-LIFE TRAPS FOLLOWING A CHANGE IN WAVELENGTH

Let us assume that a semiconductor with long-life traps was illuminated with light of wavelength λ_i. The resulting steady-state conditions will correspond to a hole concentration p_{ss} (λ_i). Let us now expose the semiconductor to illumination at another wavelength λ and intensity L. When the wavelength is thus changed, the original steady-state conditions, in which the rate of filling of empty s-centers equaled the rate of ejection of carriers from filled s-centers, is disturbed. As a result, the concentrations both of free electrons and holes and of electrons in traps begin to change. These changes of concentration will continue until a new steady state is set up, with a new value of the free hole concentration, which will also correspond to a new steady-state value of the anomalous photoconductivity ("erasing the memory"). Since $\tau(\lambda)$, the time for the establishment of the new equilibrium, is large compared with τ_0, the lifetime of free carriers (in the experiments, $\tau(\lambda)$ varied from some tenths of a second to many seconds, depending on the illumination conditions, whereas $\tau_0 \simeq 10^{-7}$ sec), the establishment of the equilibrium state can be separated into two stages. During the first stage, which takes a few τ_0 $(\simeq 5\tau_0)$, the concentration of free electrons reaches a value close to the steady-state concentration $(n \simeq n_{ss})$, while the concentration of electrons in traps remains practically unchanged, equal to q_{ss}. Consequently, during this stage

$$\frac{dp}{dt} = \frac{dn}{dt} = K\beta L - \frac{n}{\tau_0} . \tag{2.18}$$

The solution of (2.18) is*

$$n = n_{ss}(1 - e^{-t/\tau_0}) = K\beta\tau_0 L\,(1 - e^{-t/\tau_0}) \tag{2.19}$$

so that

$$\Delta\sigma = e\,(u_p + u_n)\,K\beta L\tau_0\,(1 - e^{-t/\tau_0}). \tag{2.20}$$

However, since $K\beta\tau_0 L = n_{ss}$ is small (condition (2.17) is assumed), the overall change in conductivity during the first stage, $\Delta\sigma$, is negligible.

During the second stage $(t \gg 5\tau_0)$, the concentration of electrons in the conduction band changes very little and can be assumed to remain constant, equal to $n_{ss} = K\beta\tau_0 L$. In the course of this stage, which lasts until the steady state is established, the free electrons created by light are gradually captured by long-life traps. As a result, the concentration of free holes changes and the conductivity of the sample is thus also affected.

---

* It is assumed that all the electrons in the conduction band have recombined with holes in the valence band after the illumination is switched off, therefore $n = 0$ at $t = 0$.

For the second stage, we have

$$\frac{dp}{dt} = \frac{dq}{dt} = \gamma Q \cdot n_{ss} - SLp =$$
$$= K\beta\tau_0\gamma LQ - SLp = K\beta'L - SLp, \qquad (2.21)$$

where

$$\beta' = \beta\tau_0\gamma Q \qquad (2.22)$$

is the quantum yield of the process in which a photoelectron hits a long-life trap.

The solution of equation (2.21) can be written in the form

$$p_{ss}(\lambda) - p = [p_{ss}(\lambda) - p_{ss}(\lambda_i)]\, e^{-SLt}, \qquad (2.23)$$

and the corresponding conductivity is

$$\sigma_s - \sigma = (\sigma_s - \sigma^{(0)})\, e^{-SLt}. \qquad (2.24)$$

Since the change of the free hole concentration (and conductivity) during the first stage is small, the initial values of these quantities in (2.23) and (2.24) are taken as their value at the time $t = 0$.

During the initial phase of the second stage, i. e., for a time t not much larger than $5\tau_0$, the rate of change of the hole concentration is constant and equal to

$$\frac{dp}{dt} = K\beta'L - Sp_{ss}(\lambda_i)\, L. \qquad (2.25)$$

The solution of (2.25) is

$$p - p_{ss}(\lambda_i) = (K\beta'L - Sp_{ss}(\lambda_i)\, L)\, t \qquad (2.26)$$

and correspondingly

$$\sigma - \sigma^{(0)} = eu_p\, (K\beta' - Sp_{ss}(\lambda_i))\, Lt. \qquad (2.27)$$

The general shape of the curve $\Delta\sigma = \Delta\sigma\,(t)$ is shown in Figure 12. The initial section of this curve is presented in Figure 12b on an enlarged scale (the time scale is stretched disproportionately). From the initial section of $\Delta\sigma\,(t)$, using the angles φ_1 and φ_2 (Figure 12a, b), we can determine the values of β and β'. On Figure 4, which gives the relaxation curve of the conductivity following a change in wavelength, the initial section does not appear because n_{ss} and τ_0 are too small as a result of the relatively low light intensity. Therefore, the shape of the curves observed on the oscillograms fully conform to relation (2.24), i.e., when condition (2.17) is fulfilled, the experimental conductivity relaxation curves are exponential.

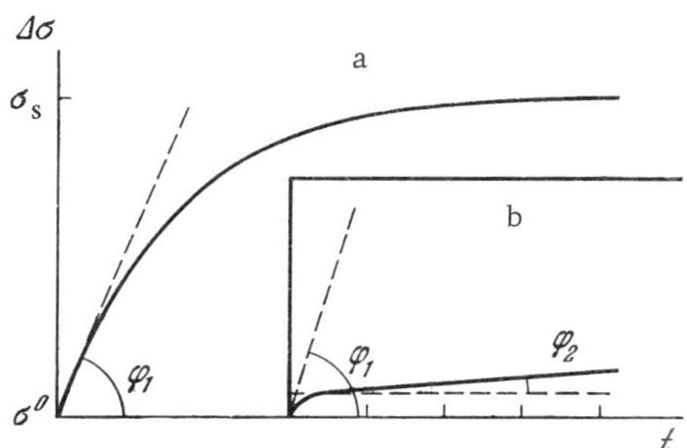

FIGURE 12. The kinetics of the establishment of photo-conductivity in a semiconductor with long-life traps

Further, it follows from (2.24) that the conductivity relaxation time $\tau(\lambda)$ for illumination at wavelength λ and intensity L is

$$\tau(\lambda) = \frac{1}{SL}, \qquad (2.28)$$

i.e., it does not depend on the conditions of previous illumination of the sample (λ_i, L_i), but only on the properties of the current illumination (items 10 and 11 in the account of properties of anomalous photoconductivity).

It also follows from (2.24) that the derivative, calculated from the experimental curve, is given by the relation

$$\left(\frac{d\sigma}{dt}\right)^{\lambda_i \to \lambda}_{t \to 0} = [\sigma_s(\lambda) - \sigma_s(\lambda_i)]\,SL = \frac{\sigma_s(\lambda) - \sigma_s(\lambda_i)}{\tau(\lambda)}. \qquad (2.29)$$

For the reverse change in wavelength from λ to λ_i, we have

$$\left(\frac{d\sigma}{dt}\right)^{\lambda \to \lambda_i}_{t \to 0} = \frac{\sigma_s(\lambda_i) - \sigma_s(\lambda)}{\tau(\lambda_i)}. \qquad (2.30)$$

Since $\tau(\lambda) \neq \tau(\lambda_i)$, it follows from (2.29) and (2.30) that $\left(\frac{d\sigma}{dt}\right)_{t \to 0}$ has a different sign for these changes of wavelength and differs in absolute value (item 6 in the account of properties of anomalous photo-conductivity).

Furthermore, since $\sigma_s\,(\lambda)$ increases with wavelength in the visible spectrum, $\left(\dfrac{d\sigma}{dt}\right)_{l\to 0}^{\lambda_i\to\lambda}$ is positive for $\lambda > \lambda_i$ (positive photo-conductivity), and negative for $\lambda < \lambda_i$ (negative photoconductivity).

Finally, the last point. It follows from (2.28) that

$$\tau\,(\lambda)\cdot L = f\,(\lambda), \tag{2.31}$$

i.e., for radiation of a given spectral composition, the product $\tau\,(\lambda)\cdot L$ is constant, independent of light intensity.

It follows from Secs. 2 and 3 of this chapter that a semiconductor containing only one type of impurity centers, i.e., s-centers, possesses all the properties of anomalously photoconductive amorphous selenium enumerated in Chapter 1 when condition (2.17) is fulfilled.

Nothing can be said in the framework of the phenomenological theory as regards the dependence of $\tau\,(\lambda)$ on λ. The function $\tau\,(\lambda)$ is determined by the structure and the properties of the long-life traps and it cannot be found unless the nature of the traps is clarified and a suitable microscopic theory is constructed.

4. PHOTOCONDUCTIVITY ASSOCIATED WITH THE DIRECT CAPTURE OF ELECTRONS FROM THE VALENCE BAND INTO LONG-LIFE TRAPS

When a semiconductor is illuminated with wavelengths from its intrinsic absorption region, the capture of electrons by long-life traps from the conduction band may turn out to be negligible compared with the direct excitation of electrons from the valence band. Let us examine what properties the photoconductivity should have in this case /16/.

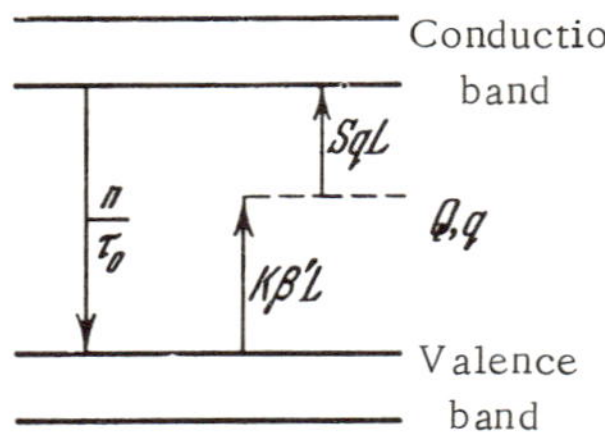

FIGURE 13. Electron transitions scheme in a semiconductor containing only long-life traps: direct excitation by photons of electrons from the valence band into long-life traps

The corresponding electron transition scheme is given in Figure 13. No transitions from the conduction band to the long-life trap are shown because it is assumed that there are no electrons in the conduction band (illumination with photon energy ε smaller than the forbidden gap width), and that the main mechanism of electron capture by long-life traps is the direct excitation of an electron from the valence band by a photon.

The kinetic equations corresponding to this scheme have the form

$$\frac{dn}{dt} = SqL - \frac{n}{\tau_0} , \qquad (2.32)$$

$$\frac{dq}{dt} = K\beta'L - SqL, \qquad (2.33)$$

$$p = n + q. \qquad (2.34)$$

The notation n, S, β', p, K and q used in (2.32)–(2.34) has the same meaning as in the previous discussion.

It follows from these equations that the rate of filling of the s-centers with carriers does not depend on the concentration q. The concentration n is established quickly because τ_0 is small and, therefore, n will follow the changes in the concentration q and, in the steady state, attain the value

$$n_{ss} = Sq_{ss}L\tau_0. \qquad (2.35)$$

If

$$SL\tau_0 \ll 1, \qquad (2.36)$$

then

$$n_{ss} \ll q_{ss}, \qquad (2.37)$$

so that

$$p = n + q \simeq q. \qquad (2.38)$$

If $p \simeq q$, the kinetics of the changes in p will be described by an equation analogous to (2.33)

$$\frac{dp}{dt} = K\beta'L - SpL, \qquad (2.39)$$

which coincides with equation (2.25).

The quantity β' has the same physical meaning both for capture of electrons by long-life traps from the conduction band and for excitation from the valence band. It represents the probability that an electron in selenium will hit a long-life trap after absorbing a single photon.

Since equations (2.25) and (2.39) coincide, all the conductivity properties of a semiconductor with long-life traps, established in the region $\lambda < \lambda_{ed}$ (λ_{ed} is the absorption edge), remain valid in the spectral region $\lambda > \lambda_{ed}$, inside the absorption band; in particular the steady-state conductivity is independent of the light intensity, and the spectral memory must show up, together with all the other properties of anomalous photoconductivity.

5. PHOTOCONDUCTIVITY OF A SEMI-CONDUCTOR CONTAINING S-CENTERS AND ORDINARY CENTERS

If, in addition to s-centers, the semiconductor also contains other centers (recombination centers or trapping levels), the character of its photoconductivity will naturally change, and it will have features other than those described in Secs. 1—3 of the present chapter. Does such a semiconductor exhibit anomalous photoconductivity inherent in a semiconductor with s-centers only, and if it does, under what conditions?

The photoconductivity of a semiconductor containing long-life traps and centers of other types was investigated elsewhere /18/. The fundamental results depend little on the nature of the other centers, and we shall therefore restrict our discussion to a single case, namely a semiconductor with s-centers and centers inter-acting with the valence band. When such a semiconductor is illuminated, the generation of free carriers is accompanied by a charge redistribution among the impurity centers.

Figure 14 shows the charge (electron) transitions scheme relevant to this case. It is assumed that the s-centers capture electrons, i.e., the semiconductor has hole-type (p-type) conductivity.

In Figure 14 the dashed lines represent both the long-life traps (whose concentration is Q) and the centers interacting with the valence band (trapping centers, concentration M). Filled centers (centers which have captured an electron) are represented in Figure 14 by circled dashes; their respective concentrations are q (filled long-life traps) and m (filled trapping levels). As in

Sec. 1 of the present chapter, we assume that free electrons reach the long-life traps through the conduction band and that electrons are liberated from long-life traps only by absorbing a photon. We shall also assume, as before, zero cross-section for the recombination of a free hole with an electron bound to a long-life trap and zero probability for thermal excitation of an electron from a long-life trap into the conduction band.

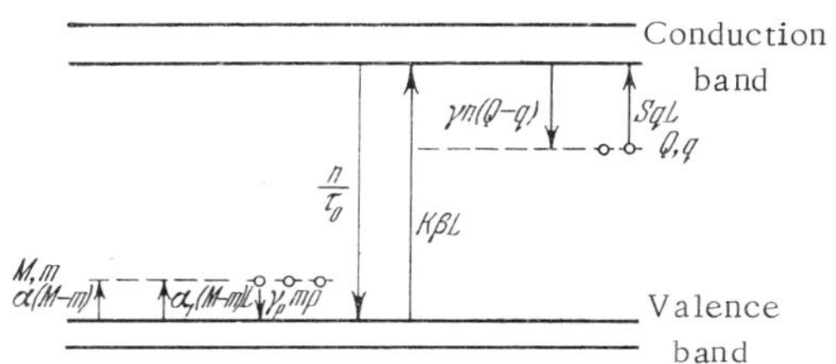

FIGURE 14. Transition scheme in a semiconductor containing long-life traps (s-centers) and centers interacting with the valence band

The trapping levels (their concentration is M) capture electrons (generate free holes) both by thermal excitation and by absorption of photons. The rate of thermal hole generation is proportional to the concentration of the empty trapping centers $(M - m)$. The proportionality coefficient α is determined by the effective density of states in the valence band. The rate of photogeneration is proportional to the concentration of the empty impurity centers and to the light intensity L. The number of free holes captured per second by negatively charged trapping centers is $\gamma_p mp$, where γ_p is the per-second capture probability of a free hole by one of the charged trapping centers (the probability for the transition of an electron from a trapping center into the valence band).

The kinetic equations for the relaxation of photoconductivity can be written in the following form under our assumptions:

$$p = q + m + n, \tag{2.40}$$

$$\frac{dn}{dt} = K\beta L - \frac{n}{\tau_0} - n\gamma (Q - q) + SqL, \tag{2.41}$$

$$\frac{dq}{dt} = n\gamma (Q - q) - SqL, \tag{2.42}$$

$$\frac{dm}{dt} = \alpha (M - m) + \alpha_1 L (M - m) - \gamma_p mp, \tag{2.43}$$

where, as before, S is the cross-section for the photoexcitation of an electron from a long-life trap, γ is the probability of the trap capturing an excess electron, α and α_1 are constants representing the capture probability of an electron by a trapping center due to thermal motion or photon absorption, respectively.

It follows from (2.40) that q, the concentration of filled long-life traps, does not depend directly on M or m, but is determined only by the number of electrons in the conduction band. If τ_0, the lifetime of free electrons and holes, is much less than all the other relaxation times, as is the case in amorphous selenium ($\tau_0 \simeq 10^{-7}$ sec, all the other relaxation times being a second or longer), the concentration of electrons in the conduction band is just as in the case when it was assumed that there were only valence centers in amorphous selenium) established in a very short time ($\sim 5\,\tau_0$). Therefore, as for a semiconductor with s-centers only, we can assume this equilibrium is established instantaneously when illumination is switched on.

The steady-state concentration of free electrons n_{ss} established by illumination is determined from the relation

$$K\beta L - \frac{n_{ss}}{\tau_0} = 0, \qquad n_{ss} = K\beta L\tau_0. \tag{2.44}$$

Using (2.44), we rewrite (2.42), which determines the concentration of filled long-life traps, in the form

$$\frac{dq}{dt} = K\beta L\tau_0\gamma\,(Q - q) - SqL. \tag{2.45}$$

It follows from (2.45) that, under the above assumptions, the influence of impurity centers (of any type) on the filling of long-life traps and thus also on the anomalously photoconducting properties is entirely determined by the effect these centers have on the magnitude of τ_0. Both σ_a and σ_n increase with increasing τ_0. However, the basic features of anomalous photoconductivity, namely the conservation of σ_a in the dark and the constancy of σ_a at various light intensities, are retained in this case also. $\sigma_n = K\beta L\tau_0$ increases with τ_0, and for sufficiently large τ_0, σ_n will no longer be negligibly small in comparison with σ_a.

When the wavelength is changed, the relaxation of σ_a is determined by the change in the concentration q, as expressed by equation (2.46) analogous to equation (2.23),

$$q_{ss} - q = (q_{ss} - q_0)\,e^{-SLt}, \tag{2.46}$$

where q_{ss} is the steady-state concentration of filled long-life traps, established in the sample after sufficiently prolonged illumination. q_{ss} is given by

$$q_{ss} = \frac{K\beta\tau_0\gamma}{S + K\beta\tau_0\gamma}\, Q.$$

When the condition of anomalous photoconductivity

$$K\beta\tau_0\gamma \ll S$$

is fulfilled, q_{ss} is given by

$$q_{ss} \simeq \frac{K\beta\tau_0\gamma}{S}\, Q = \frac{K\beta'}{S}, \tag{2.47}$$

i. e., it is expressed by the same equation as for a semiconductor containing s-centers only.

However, the full conductivity of the sample is now determined by a combination of factors and not the concentration of filled long-life traps alone.

Since the free hole concentration is $p = q + m$ when $n \ll q$ (the anomaly condition), the observed conductivity σ is no longer equal to $eu_p q$ and is given by

$$\sigma = eu_p\,(q + m). \tag{2.48}$$

The steady-state q and its instantaneous value during relaxation following a change in wavelength do not depend on m, so that the conductivity of the sample

$$\sigma = eu_p\,(q + m) = eu_p q + eu_p m = \sigma_a + \sigma_n \tag{2.49}$$

can be written as a sum of two independent terms, one of which represents the contribution of carriers associated with the s-centers in the semiconductor (anomalous photoconductivity), and the other the contribution of carriers associated with M centers. The conductivity behaves in such a way as if governed by two independent photoconductivity mechanisms.

In order to establish the character of relaxation and determine the steady-state conductivity under illumination (after q has been determined by (2.46) and (2.47)), it is now sufficient to determine m, i. e., to solve equation (2.43) which, using relation (2.40) and the condition for the existence of anomalous photoconductivity $n \ll q$, takes the form

$$\frac{dm}{dt} = (\varkappa + \alpha_1 L)(M - m) - \gamma_p(q + m)\, m. \tag{2.50}$$

Since q is an exponential function of time (2.46), it is difficult to determine the character of the relaxation of m under illumination; the steady-state value under illumination m_{ss} can be found by solving the equation

$$(\varkappa + \alpha_1 L)\, M - (\alpha + \alpha_1 L + \gamma_p q_{ss})\, m_{ss} - \gamma_p m_{ss}^2. \tag{2.51}$$

It follows from (2.51) that

$$m_{ss} = -\frac{1}{2}\left(\frac{\alpha + \alpha_1 L}{\gamma_p} + q_{ss}\right) + \sqrt{\frac{1}{4}\left(\frac{\alpha + \alpha_1 L}{\gamma_p} + q_{ss}\right)^2 + \frac{\alpha + \alpha_1 L}{\gamma_p}\, M}. \tag{2.52}$$

Let us consider two cases:

a) The concentration of impurity centers M is large. In this case

$$\left(\frac{\alpha + \alpha_1 L}{\gamma_p}\, M\right)^{1/2} \gg \frac{1}{2}\left(\frac{\alpha + \alpha_1 L}{\gamma_p} + q_{ss}\right). \tag{2.53}$$

Hence,

$$m_{ss} \simeq \sqrt{\frac{\alpha + \alpha_1 L}{\gamma_p}\, M} \tag{2.54}$$

and

$$\sigma_s = e u_p\left(q_{ss} + \sqrt{\frac{\alpha + \alpha_1 L}{\gamma_p}\, M}\right). \tag{2.54a}$$

It follows from (2.53) that q_{ss} is much smaller than $\left(\dfrac{\alpha + \alpha_1 L}{\gamma_p}\, M\right)^{1/2}$, so that

$$\sigma_s = e u_p \sqrt{\frac{\alpha + \alpha_1 L}{\gamma_p}\, M}. \tag{2.55}$$

The photoconductivity behaves as if there were no long-life traps in the sample. It is clear that the case under consideration cannot occur in a sample exhibiting anomalous photoconductivity.

b) The concentration of impurity centers M is small. In this case,

$$\frac{\alpha + \alpha_1 L}{\gamma_p} M \ll \frac{1}{4}\left(\frac{\alpha + \alpha_1 L}{\gamma_p} + q_{ss}\right)^2, \tag{2.56}$$

and

$$m_{ss} = \frac{\dfrac{\alpha + \alpha_1 L}{\gamma_p} M}{\dfrac{\alpha + \alpha_1 L}{\gamma_p} + q_{ss}} = \frac{(\alpha + \alpha_1 L) M}{\alpha + \alpha_1 L + \gamma_p q_{ss}}. \tag{2.57}$$

By (2.56) and (2.57)

$$m_{ss} \ll \frac{\alpha + \alpha_1 L}{\gamma_p} + q_{ss}. \tag{2.58}$$

If the impurity centers are such that $\frac{\alpha + \alpha_1 L}{\gamma_p}$ is not very large compared with q_{ss}, then m_{ss} will be very small compared with q_{ss}. This is precisely the case we are interested in. The contribution of the impurity centers M to photoconductivity in this case is small compared with the contribution of the long-life traps.

The magnitude of the steady-state conductivity under illumination σ_s is determined in this case by the relation

$$\sigma_s = eu_+\left(\frac{K\beta'}{S} + \frac{(\alpha + \alpha_1 L) M}{\alpha + \alpha_1 L + \gamma_p \dfrac{K\beta'}{S}}\right). \tag{2.59}$$

We shall now establish how the conductivity relaxes when the illumination is switched off. The concentration of filled long-life traps remains constant: it is equal to their concentration at the time when illumination is switched off. The concentration of filled impurity centers m begins to decrease from m_{ss} to m_0. The relaxation of the sample conductivity when the illumination is switched off is entirely determined by the relaxation of its nonanomalous component. By (2.57)

$$m_0 = \frac{\alpha M}{\alpha + \gamma_p q_{ss}}, \tag{2.60}$$

and the instantaneous value of m in the relaxation process is

$$m = \frac{m_0 - m_1 \dfrac{m_{ss} - m_0}{m_{ss} - m_1} e^{-t\sqrt{b^2 + 4ac}}}{1 - \dfrac{m_{ss} - m_0}{m_{ss} - m_1} e^{-t\sqrt{b^2 + 4ac}}} , \qquad (2.61)$$

where

$$m_1 = -\frac{b}{2c} - \sqrt{\frac{b^2}{4c^2} + \frac{a}{c}} , \qquad (2.62)$$

$$a = \alpha M, \qquad b = \alpha + \gamma_p q_{ss}, \qquad c = \gamma_p. \qquad (2.63)$$

By (2.60), the steady-state dark conductivity is

$$\sigma_d = e u_p \left(\frac{K\beta'}{S} + \frac{\alpha M}{\alpha + \gamma_p q_{ss}} \right) . \qquad (2.64)$$

We see from (2.64) that the steady-state dark value does not depend on the intensity of previous illumination. However, in distinction from the case considered in Secs. 2—4, the dark conductivity is determined not only by the concentration of the metastable carriers but depends also on the concentration of equilibrium carriers, and can be represented as the sum $\sigma_m + \sigma_0$, where σ_0 is the conductivity due to the equilibrium carriers associated with the impurity centers M, and σ_m is the conductivity due to metastable carriers.

Note that

$$\frac{\alpha M}{\alpha + \gamma_p q_{ss}} < \frac{(\alpha + \alpha_1 L) M}{\alpha + \alpha_1 L + \gamma_p q_{ss}} ,$$

and therefore $\dfrac{\alpha M}{\alpha + \gamma_p q_{ss}}$ is considerably smaller than the corresponding value of q_{ss}. Consequently, the magnitude of the conductivity established after the illumination is switched off is mainly determined by the value of $q_{ss} = \dfrac{K\beta'}{S}$, i. e., it depends on the wavelength.

Thus, even if the sample contains in addition to long-life traps a small quantity of other impurity centers, it can still possess color memory. The sample remembers the spectral composition of the light with which it has been illuminated. A considerable fraction of the conductivity induced in the sample is preserved after the illumination is switched off. However, in this case the memory will only be partial. The steady-state dark conductivity differs in magnitude from the steady-state conductivity under illumination. When the illumination is switched off the conductivity must change from σ_s to σ_d, due to the change in the concentration m from m_{ss} to m_0.

In this case only its steady-state dark conductivity is independent of light intensity; σ_s is a function of light intensity, though the dependence is not a very strong one.

We shall now find the magnitude $\Delta\sigma_n = \sigma_s - \sigma_d$. Note that σ_d is the steady-state value of the dark conductivity after the given illumination has been switched off. It follows from (2.59) and (2.64) that

$$\Delta\sigma_n = eu_p \frac{M\gamma_p q_{ss}\alpha_1 L}{(\alpha + \gamma_p q_{ss})(\alpha + \alpha_1 L + \gamma_p q_{ss})} . \qquad (2.65)$$

$\Delta\sigma_n$ is interpreted as the photoresponse for the nonanomalous component of the photoconductivity. $\Delta\sigma_n$ depends on the light intensity, on the wavelength (through q_{ss}), and on the temperature (α, a quantity proportional to the effective density of states m in the valence band, is a function of the temperature). The dominant form of the dependence of $\Delta\sigma_n$ on L, λ and T is determined by which of the three terms $\alpha, \alpha_1 L, \gamma_p q_{ss}$ is dominant. For a sufficiently small incident light intensity

$$\alpha_1 L \ll \alpha + \gamma_p q_{ss},$$

the magnitude

$$\Delta\sigma_n = eu_+ \frac{\alpha_1 M\gamma_p q_{ss}}{(\alpha + \gamma_p q_{ss})^2} L$$

is directly proportional to the light intensity L. If α is then much larger than $\gamma_p q_{ss}$, $\Delta\sigma_n$ will vary exponentially with temperature (decreasing with increasing temperature) and increase with λ, since q_{ss} increases with λ. If, on the other hand, α is less than $\gamma_p q_{ss}$, and the proportional dependence on light intensity is preserved, $\Delta\sigma_n$ depends little on the temperature and decreases with increasing wavelength.

When the light intensity is large, $\Delta\sigma_n$ varies with L more slowly and sublinearly, and for sufficiently large L it becomes independent of intensity.

Note that the photoresponse $\Delta\sigma_n$ due to the presence of M centers in the semiconductor can be written also in the form

$$\Delta\sigma_n = K\beta L\tau_n, \qquad (2.66)$$

where τ_n is the lifetime of the majority excess carriers associated with the M centers in the semiconductor. The steady-state conductivity under illumination σ_s is written in the form

$$\sigma_s = \sigma_a(\lambda) + \Delta\sigma_n + \sigma_0. \qquad (2.67)$$

After the illumination is switched off, $\Delta\sigma_n$ falls to zero with relaxation time τ_n. $\Delta\sigma_n$ constitutes the conductivity component associated with carriers which are generated in the semiconductor in the dark as a result of the presence of impurity trapping centers at a concentration M.

The conductivity relaxation (2.61) in our case is rather complex. However, it will behave in the main as if there were no long-life traps in the sample.

Indeed, let Q (and consequently also q) be zero. Equation (2.61) is transformed into (2.68),

$$\frac{m_d - m}{m_1 - m} = \frac{m_s - m_d}{m_1 - m_d}\, e^{-t\sqrt{b_1^2 + 4ac}}. \qquad (2.68)$$

The quantities which enter (2.68) are as follows:

$$b_1 = \alpha; \quad m_1 = -\frac{b_1}{2c} - \sqrt{\frac{b_1^2}{4c^2} + \frac{a}{c}}. \qquad (2.69)$$

A comparison of relations (2.68) and (2.69), which determine the steady-state conductivity under illumination and the relaxation curves for a semiconductor with M centers only, with relations (2.51) and (2.63), which determine the steady-state value of the anomalous component of conductivity and the relaxation curves for a semiconductor with M centers and long-life traps Q, shows that, although $\Delta\sigma$ may differ considerably from the conductivity under illumination in a semiconductor with no long-life traps, the relaxation curves have the same character, differing only in the coefficients b and m_1.

6. IMPLICATIONS OF THE PHENOMENOLOGICAL THEORY OF ANOMALOUS PHOTOCONDUCTIVITY

We shall now consider some implications which follow from the phenomenological theory.

1. Let us clarify between what limits anomalous photoconductivity should exist according to this theory, and under what conditions it disappears. In a semiconductor with s-centers, the steady-state photoconductivity does not depend on light intensity and is retained after the illumination is switched off. In a semiconductor containing in addition to s-centers a small quantity of other centers the photo-conductivity consists of two components: 1) an anomalous component $\sigma_a(\lambda) = eu_p \frac{K\beta'}{S}$, which does not depend on light intensity and is pre-served after the removal of illumination, and 2) a normal component with the photoresponse $\Delta\sigma_n = eu_p K\beta L\tau_n$, which decays with relaxation time τ_n when the illumination is switched off, so that the dark value of conductivity, irrespective of the light intensity, is $\sigma_a + \sigma_0$, where σ_0 is the equilibrium value of the nonanomalous component.

Consequently, the main characteristics of anomalous photocon-ductivity are that $\sigma_a(\lambda)$ is independent of L and that its value is preserved after the removal of illumination. A deviation from the properties of anomalous photoconductivity will therefore manifest itself as a violation of these basic features. The independence between $\sigma_a(\lambda)$ and L in the phenomenological theory is due to two factors: a) a carrier captured in a s-center remains there for an indefinitely long time; in other words, the probability of spontaneous excitation of a captured carrier from an s-center is assumed to be zero; 2) the s-centers are filled only by excess carriers; in other words, the concentration of minority equilibrium carriers is assumed to be negligibly small.

The non-observance of these two assumptions leads to the violation of the anomalous properties of photoconductivity. In reality, the probability of spontaneous ejection of a carrier from an s-center is not zero, although it is very small. We shall denote this probability by θ. The concentration of equilibrium carriers will be denoted by n_0. The concentration n_0 is very small, but not zero. The kinetic equation which determines the degree of filling of s-centers has the form

$$\frac{dq}{dt} = K\beta'L - SqL - \theta q + \gamma n_0(Q - q), \qquad (2.70)$$

whence it follows that the steady-state value $q = q_{ss}$, determinable from the condition

$$\frac{dq}{dt} = 0,$$

will be expressed by the relation

$$q_{ss} = \frac{\gamma (n_0 + K\beta L\tau_0) Q}{\gamma (n_0 + K\beta L\tau_0) + \theta + SL} = \frac{1 + \dfrac{n_0}{K\beta L\tau_0}}{1 + \dfrac{\theta + \gamma n_0 + SL}{K\beta L\tau_0 \gamma}} . \tag{2.71}$$

It is seen from (2.71) that q_{ss} in this case is a function of the light intensity L, and only when

$$n_0 \ll K\beta L\tau_0 \quad \text{and} \quad \frac{\theta + \gamma n_0 + SL}{K\beta L\tau_0 \gamma} \ll 1$$

can anomalous photoconductivity occur. At temperatures of 100–200°K, the above inequalities are fulfilled in amorphous selenium films activated by mercury. At room temperature and close to it, when the photoconductivity acquires a normal character, these inequalities are not fulfilled. This problem will be considered in detail in Chapters 3 and 4.

Note that the breakdown of anomalous photoconductivity may have a twofold character. After the illumination is switched off, the steady-state concentration of filled s-centers may decrease (the traps are spontaneously discharged). This will occur when $\theta q > \gamma n_0 (Q - q)$. If, however, $\theta q < \gamma n_0 (Q - q)$, the s-centers will continue to be filled even after the illumination is switched off.

2. Let us consider another result of the phenomonological theory of anomalous photoconductivity. It can be shown that if, in addition to illumination with monochromatic light of variable intensity, the anomalous photoconductor is simultaneously illuminated with background light of constant intensity but a different wavelength, the steady-state conductivity will no longer be independent of light intensity.

Let a semiconductor containing long-life traps be illuminated at two wavelengths λ_1 and λ_2 with quantum flux intensities L_1 and L_2, respectively. We shall assume that condition (2.17) — the necessary condition of anomalous photoconductivity — is fulfilled. The kinetics of conductivity changes in this case is determined by the relation

$$\frac{1}{eu_p} \frac{d\sigma}{dt} = K_1\beta_1' L_1 + K_2\beta_2' L_2 - (S_1 L_1 + S_2 L_2)\frac{\sigma}{eu_p}, \tag{2.72}$$

where β_1' and β_2' are quantum yields, S_1 and S_2 are cross-sections for electron excitation from traps at the two wavelengths.

The solution of (2.72) gives an expression for the steady-state anomalous photoconductivity $\sigma_a\,(\lambda_1,\ \lambda_2)$

$$\sigma_a\,(\lambda_1,\lambda_2)=eu_p\,\frac{K_1\beta_1'L_1+K_2\beta_2'L_2}{S_1L_1+S_2L_2}=\sigma_a\,(\lambda_1)\,\frac{1+\dfrac{K_2\beta_2'}{K_1\beta_1'}\cdot\dfrac{L_2}{L_1}}{1+\dfrac{S_2}{S_1}\cdot\dfrac{L_2}{L_1}}. \qquad (2.73)$$

Here $\sigma_a\,(\lambda_1)$ is the steady-state conductivity of the anomalous photoconductor when it is illuminated with light of wavelengths λ_1, and $\sigma_a\,(\lambda_1,\ \lambda_2)$ is the steady-state conductivity for illumination with both light fluxes.

If the intensities of the light fluxes are changed in such a way that their ratio (i. e., spectral composition) remains constant, $\sigma_a\,(\lambda_1,\ \lambda_2)$ will remain independent of the light flux intensities. However, if the ratio of the component intensities L_2/L_1 is not preserved, and since in the general case we have

$$\frac{K_2\beta_2}{K_1\beta_1}\neq\frac{S_2}{S_1}\ ,$$

the steady-state conductivity $\sigma_a\,(\lambda_1,\ \lambda_2)$ will depend on the ratio of the component intensities. When only the intensity of one flux component changes, while the other component provides a constant background illumination, $\sigma_a\,(\lambda_1,\ \lambda_2)$, being a function of the ratio of the component intensities, will depend on the intensity of the varying light flux component.

Relation (2.73) remains valid also when the incident light flux does not consist of two monochromatic fluxes but of fluxes extending over two fairly wide spectral bands. If the ratio of their intensities is constant, the steady-state conductivity σ_s will again be independent of light intensity. If, however, the intensities of the light flux components vary disproportionately (so that the spectral composition of the radiation changes), σ_s will vary with the ratio L_2/L_1.

To prove this corollary of the phenomenological theory of anomalous photoconductivity, the current-light intensity characteristic of an anomalously photoconductive amorphous selenium sample simultaneously illuminated with light fluxes of $\lambda_1 = 650\,\text{nm}$ and $\lambda_2 = 450\,\text{nm}$ was investigated in /19/. The intensity of the second light flux L_2 was kept constant (background illumination), while the intensity of the first L_1 was varied. Figure 15a shows the observed dependence of σ_s on L_1. The solid line 3 is constructed according to calculations from (2.73), the points representing the experimental results. In addition, the straight lines 1 and 2 in Figure 15a plot the steady-state conductivity for the monochromatic light fluxes $\lambda_1 = 650\,\text{nm}$ and $\lambda_2 = 450\,\text{nm}$, respectively. Figure 15b shows the calculated form of σ_s as a function of L_2 and the experimental points for constant long-wave background illumination ($\lambda_1 = 650\,\text{nm}$) and varying intensity L_2 of the short-wave component. The measurements with short-wave and long-wave background illuminations were carried out using different samples.

It is seen that the experimental points closely follow the calculated curves, and consequently the corollary under consideration is indeed correct.

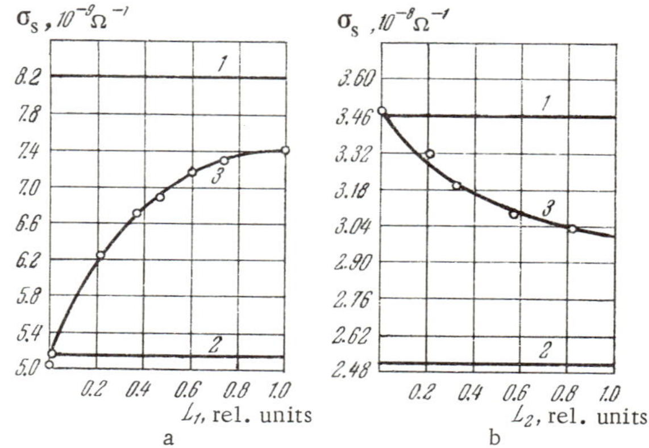

FIGURE 15. Steady-state photoconductivity vs. light intensity in anomalously photoconductive selenium exposed to a varying and constant background illumination

Note that the correlation between the conductivity in the dark and under illumination is also preserved with mixed light fluxes. The dark conductivity is still a function of the intensity of the varying light component in previous illumination.

It must be emphasized that changing the intensity ratio of the light flux components means changing the spectral composition, i.e., in this case also, the change in the steady-state conductivity is a result of a change in the spectral composition of radiation. When the spectral composition of radiation remains unchanged ($L_1/L_2 = \text{const}$), the steady-state conductivity will be independent of intensity (as for a monochromatic flux).

Let us sum up the main points considered in Chapter 2.

1. A semiconductor with a low intrinsic conductivity, containing only s-centers and satisfying condition $\gamma K \beta \tau_0 L < SL \ll \gamma Q$, will

possess all the properties which in Chapter 1 were assigned to anomalous photoconductivity.

2. A semiconductor containing in addition to s-centers a small number of other centers (recombination or trapping centers) will exhibit anomalous photoconductivity with some contribution from a nonanomalous component σ_n.

3. In the general theory of photoconductivity, in addition to recombination centers (through which carriers recombine) and trapping levels (from which carriers are thermally excited into a band), we have to consider also s-centers, whose main role is generation of metastable carriers.

Since the anomalous photoconductivity observed in films of mercury-activated amorphous selenium can be explained by the presence of s-centers, it is natural to try to clarify the properties which these centers possess and their possible nature. These problems will be considered in Chapter 3.

Chapter 3

SOME PROPERTIES OF LONG-LIFE TRAPS

1. THE SPATIAL DISTRIBUTION OF LONG-LIFE TRAPS IN ANOMALOUSLY PHOTO-CONDUCTIVE AMORPHOUS SELENIUM LAYERS

When one investigates the properties of long-life traps and, in particular, in determining the dependence of the cross-section for the excitation of a carrier from a trap on the quantum energy (which will be considered in the following section), we require the distribution of s-centers across the selenium layer. Two types of such distributions are possible.

1) The s-centers are concentrated in a layer close to the surface.

2) The s-centers are somehow distributed throughout the bulk of the layer. One of the particular cases of such a distribution is the uniform distribution. The assumption of a uniform distribution of long-life traps across the selenium layer was used in a previous publication /20/.

To find out which of the two possibilities mentioned is in fact realized in layers of anomalously photoconductive amorphous selenium, the kinetics of the establishment of the steady-state photoconductivity was investigated at different wavelengths, both for illumination of the sample on the side of the layer and on the side of the substrate /21/. What can be expected from these experiments?

Let us assume that the sample is illuminated with radiation of $\lambda = 740\,\mathrm{nm}$, weakly absorbed in selenium. This wavelength is relatively little absorbed in a selenium layer less than two microns thick, and the sample can be considered almost uniformly illuminated. The establishment of the steady-state anomalous photoconductivity will then be identical for the illumination of the sample both on the side of the layer and on the side of the substrate, irrespective of the particular distribution of long-life traps in the layer. When the steady-state value is attained, the concentration of charged long-life traps which have captured a free current carrier is equal to $q_{ss}(740)$. The experiments confirmed this expected effect. In all the samples, the kinetics of the establishment of conductivity for red-light illumination was the same whether the samples were illuminated on the side of the layer or on the side of the substrate.

Let us further assume that following the establishment in the sample of steady-state conductivity corresponding to $\lambda = 740\,\text{nm}$, the sample is illuminated with light of wavelength $\lambda = 420\,\text{nm}$. This wavelength is very readily absorbed in amorphous selenium. A selenium layer $0.035\,\mu\text{m}$ thick reduces this radiation by one half. It is evident that in this case the incident light intensity varies markedly across the sample. At a depth of $0.35\,\mu\text{m}$, it is a thousand times less than at the surface. Since $q_{ss}(740)$ is significantly larger than $q_{ss}(420)$, the $\lambda = 420\,\text{nm}$ illumination, switched on after the illumination of the sample at $\lambda = 740\,\text{nm}$, will largely discharge the long-life traps filled during the red-light exposure. The change in concentration according to the phenomenological theory is expressed in this case by

$$\frac{dq}{dt} = K\beta'L - SqL = Sq_{ss}(420)\,L - SqL, \qquad (3.1)$$

where S is the cross-section for the excitation of a carrier from a trap at $\lambda = 420\,\text{nm}$. By (3.1)

$$q - q_{ss}(420) = \left[q_{ss}(740) - q_{ss}(420)\right] e^{-SLt}. \qquad (3.2)$$

It follows from (3.2) that when the sample is illuminated for a short time Δt, only some of the traps manage to recharge. In the surface layer, the degree of recharging of traps is close to unity, but in deeper layers the degree of recharging will differ considerably from unity: the deeper the layer, the smaller the proportion of traps which manage to recharge during the time Δt.

The thickness of the layer in which the concentration of charged traps reaches the value $q(420)$, corresponding to the steady-state value for $\lambda = 420\,\text{nm}$, naturally depends on the duration of illumination with this wavelength. For illumination times used in the investigations (up to $50\,\text{min}$), the discharge of traps at depths greater than $0.35\,\mu\text{m}$ is so small that it can be neglected.

If the traps are distributed across the entire sample, then after illumination at $\lambda = 420\,\text{nm}$, the concentration of traps in the layer close to the surface will be $q_{ss}(420)$, in the layer adjacent to the substrate $q_{ss}(740)$ and between these two layers q_{ss} will have an intermediate value. If the traps were indeed distributed across the whole sample, the conductivity established during a definite period of illumination at $\lambda = 420$ nm would depend on the nature of the light with which the sample had been illuminated earlier, for in that part of the sample where the $\lambda = 420$ nm light does not penetrate, the concentration of filled long-life traps created by the previous illumination is preserved. Since it has been shown that the conductivity established in a sample illuminated with short-wave radiation also depends on previous history, i. e., on the wavelength of its earlier illumination, one must expect a nonuniform distribution of long-life traps across the sample cross-section. The long-life traps are mainly to be found in a layer close to the surface.

More detailed information on the location of long-life traps is given by the following experiments /21/:

1. Anomalously photoconductive selenium previously illuminated with red light $(\lambda = 740$ nm$)$ was illuminated on the layer side with $\lambda = 420$ nm. This illumination proved sufficient for the establishment of an anomalous photoconductivity value characteristic of $\lambda = 420$ nm across the entire sample. One can verify this by illuminating the sample again with the same wavelength but on the substrate side. Control experiments with such illumination showed that the magnitude of anomalous photoconductivity did not change.

Since $\lambda = 420$ nm light is absorbed in a layer much thinner than the full thickness of the sample $(0.8-1\,\mu$m$)$, the experiment confirms that the long-life traps are distributed in a layer close to the surface, and the rest of the sample contains no long-life traps, or their quantity is so small that the influence on conductivity is negligible.

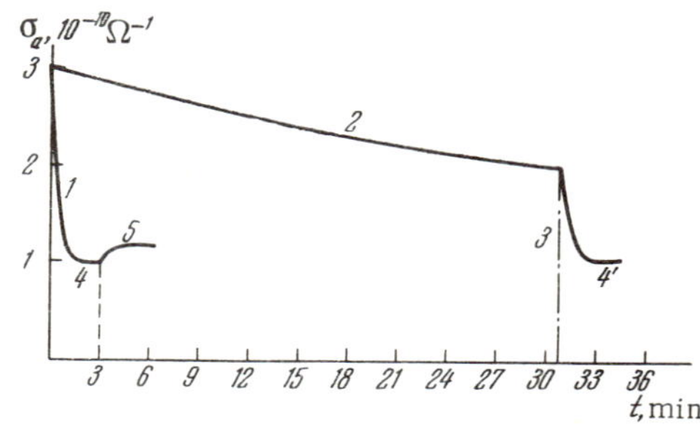

FIGURE 16. Conductivity relaxation curves of anomalously photoconductive selenium when illuminated with 420 nm light

2. The second experiment was similar to the foregoing with the difference that after the preliminary red-light illumination the sample was illuminated at $\lambda = 420$ nm not on the layer side but on the substrate side. Since the long-life traps are concentrated in a layer close to the surface, illumination with blue light on the substrate

side was not expected to cause a notable change in conductivity. This, however, was not so. A change in conductivity occurred in this case also.

Figure 16 shows relaxation curves obtained with one of the anomalously photoconductive selenium samples /21/. Curve 1 represents the relaxation of conductivity at $\lambda = 420\,\text{nm}$ when the sample is illuminated on the selenium layer side. For the given light intensity ($L = 10^{-4}\,\text{W/cm}^2$), the steady-state conductivity $\sigma_a(420)$ corresponding to $\lambda = 420\,\text{nm}$ is fully established in $\simeq 1.5\,\text{min}$ (curve 4). After the light is switched off, a steady-state level of dark conductivity is established (curve 5). Curve 2 represents the conductivity relaxation of a sample previously pre-exposed to red light when illuminated at $\lambda = 420\,\text{nm}$ on the substrate side. It turns out that the sample conductivity changes in this case also, but more slowly. The steady-state value $\sigma_a(420)$ takes much longer to establish. During the 30.5 min of illumination, the conductivity changed by somewhat less than half the difference $\sigma_a(740) - \sigma_a(420)$.

3. After a 30.5 min exposure, the light beam was switched to illuminate the sample on the layer side. The process of conductivity change (curve 3) was markedly accelerated and the steady-state value $\sigma_a(420)$ was established in approximately 1 min (curve 4').

The fact that blue-light illumination on the substrate side leads to a change in conductivity proves that blue light acts on the long-life traps in this illumination geometry also. The slower relaxation (than that for layer-side illumination) only indicates that, for substrate illumination, the light intensity at the site of the long-life traps is less than for illumination on the selenium layer side. It is obvious that if the sample is illuminated on the substrate side with higher intensity light, the relaxation time is reduced. For a sufficiently large ratio L_1/L, where L_1 is the illumination intensity on the substrate side and L that on the layer side, the relaxation times may be equalized. It turns out that if the sample is illuminated on the substrate side at $\lambda = 420\,\text{nm}$ with light intensity 64 times that used on the layer side, the discharge time of long-life traps will be the same in the two cases, i.e., in roughly 1.5 min the photoconductivity of the sample goes over from $\sigma_a(740)$ to $\sigma_a(420)$.

The results of these experiments confirm the above conclusion that the long-life traps are distributed in a layer close to the surface and make it possible to estimate the thickness of this layer.

Indeed, $\lambda = 420\,\text{nm}$ light is attenuated a factor of 64 times in an amorphous selenium layer 0.22 μm thick, and consequently the selenium layer containing the long-life traps is located not farther than 0.22 μm from the substrate. The thickness of the selenium

layer in the sample is 0.8—1.0 μm, which is much greater than
0.22 μm. The simultaneous requirements that the long-life traps
in the sample be located close to the surface and not farther than
0.22 μm from the substrate can be satisfied only if we assume that
the thickness of the selenium layer is not constant. The layer
must have a complex topography (Figure 17) with numerous hills
and valleys penetrating almost to the substrate to a distance of at
most 0.22 μm from the substrate. The long-life traps are distributed in the valleys of this microrelief.

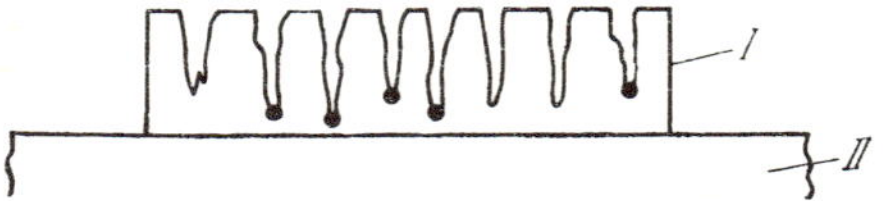

FIGURE 17. Schematic illustration of the surface topography
of a selenium layer obtained by deposition on a substrate. I is
the selenium layer, II the substrate. The black circles illustrate
the long-life traps.

The existence of a surface topography in amorphous selenium
films was noted in a previous publication /22/. Direct examination
of the surface under an electron microscope also confirmed the
presence of a microrelief and made it possible to establish its
character.

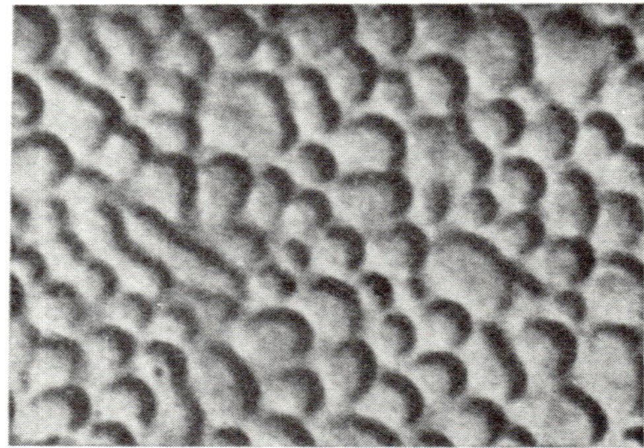

FIGURE 18. A photograph of the surface of a selenium
layer observable under an electron microscope

Figure 18 shows an electron micrograph of the surface obtained
by the method of graphite replica with platinum shadowing under a

magnification of 5500. The "granular" structure of the layer is
quite clear. The dimensions of the amorphous selenium "granules"
are within $0.2-1.5\,\mu$m. The depth of the valleys was determined
so crudely that it should be regarded as an estimate rather than a
measurement. According to this estimate, the depth is between
0.2 and $0.5\,\mu$m.

2. ESTIMATE OF THE LOWER LIMIT OF CARRIER LIFETIME IN S-CENTERS

The life-time of a trapped carrier τ_q is one of the most important characteristics of an s-center. It is the large value of τ_q
that distinguishes the s-centers from other type of centers. In the
phenomenological theory of anomalous photoconductivity, τ_q is
assumed to be infinite and it does not figure in the theoretical equations. Attempts to determine this magnitude experimentally /7, 23/
were not successful. τ_q is indeed so large that it cannot be measured and it is only possible to assess the lower limit of its probable
value. We shall present here the results of one of the numerous
experiments for the determination of τ_q. Since τ_q is large, prolonged observations of conductivity changes were necessary for its
determination. The sample was placed in a special low-temperature
vacuum cryostat /7/ with autonomous pumping. In this cryostat,
the sample temperature could be kept constant for many days. In
the course of one week, the pressure in the cryostat changed from
10^{-7} to 10^{-5} torr. The sample conductivity was measured by the
voltage drop across a standard resistance which was kept at constant temperature. The fluctuations in the readings did not exceed
0.7% during the entire session.
The experiment itself was conducted in the following manner.
The sample was illuminated at $\lambda = 420\,$nm. After the illumination
was switched off, the sample was left in the dark for some time
until a definite steady-state value σ_d was established. The sample
was then illuminated with light of a different wavelength, namely
$\lambda = 730\,$nm. As a result, the conductivity changed and stabilized at
a new steady-state level $\sigma_a(730)$.
The illumination was subsequently switched off and the conductivity relaxation in the dark was observed for 110 hours (about five
days). The experiment was conducted with sample No. 5VK at 123°K.
The observed relaxation curve is presented in Figure 19. In this
figure the abscissa gives the time and the ordinate the sample conductivity. Three curves are shown in the figure: curve II represents

the conductivity relaxation of the sample after the red light ($\lambda = 730\,nm$) is switched off. The straight line I represents the steady-state value of conductivity established in the sample when illuminated with light of $\lambda = 730\,nm$. The zero of the time axis in Figure 19 corresponds to the time the illumination is switched off. The straight line III represents the conductivity level of the sample before switching on the $\lambda = 730\,nm$ illumination. We see from the figure that the relaxation process is very drawn out. The conductivity keeps declining for 40 hours. In the next 70 hours no regular change of conductivity is observed. There are only fluctuations in σ_d within the above-mentioned 0.7%. In other words, during these 70 hours the sample conductivity remains unchanged within the margin of experimental error. This constant conductivity value was $2.5 \cdot 10^{-11}\,\Omega^{-1}$, i. e., approximately 11 times the conductivity $(2.2 \cdot 10^{-12}\,\Omega^{-1})$ established in the sample after switching off the $\lambda = 420\,nm$ illumination (line III).

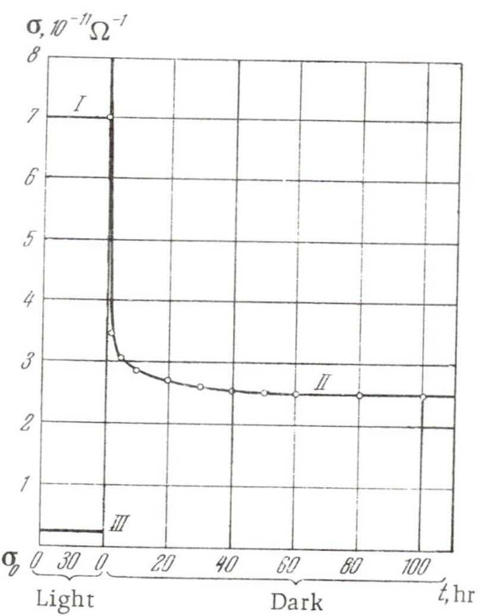

FIGURE 19. Estimating the lower limit of the carrier lifetime in an s-center

Since $\sigma_a(730)$ remains unchanged, within the experimental error, during the observation period from 40 to 110 hours after switching off the illumination, it is impossible to calculate the lifetime τ_q of a captured carrier in a long-life trap. It is, however, possible to estimate the lower limit of this lifetime.

We shall assume that in the time interval 40—110 hours after switching off the illumination the concentration of charged long-life traps continues to decrease, the traps discharge (captured carriers leave the traps), and the anomalous photoconductivity decreases, but in such a way that the change in conductivity falls within the margin of experimental error in our measurements of σ and therefore remains undetected. The error in the determination of σ in this experiment was $\pm 0.85 \cdot 10^{-13} \, \Omega^{-1}$. Consequently, the overall possible change in conductivity $\delta\sigma$ during these 70 hours could not be greater than $1.7 \cdot 10^{-13} \, \Omega^{-1}$. This conductivity decrement is not entirely due to the discharge of long-life traps, for relaxation of the nonanomalous photoconductivity component observed during the first 40 hours is also taking place in this time interval. Therefore a part of the possible change $\delta\sigma$ should be related to the relaxation of the nonanomalous component of photoconductivity. Since its variation during the first 40 hours is known, we can find an analytical expression which describes the relaxation of the nonanomalous photoconductivity component and calculate the extrapolated change during the next 40—110 hour period. It turns out that in this period the nonanomalous conductivity component must change by $\delta\sigma' = 0.6 \cdot 10^{-13} \, \Omega^{-1}$. Consequently, the anomalous photoconductivity during the relevant time changes by no more than $\delta\sigma_0 = \delta\sigma - \delta\sigma' = 1.1 \cdot 10^{-13} \Omega^{-1}$, i. e., somewhat less than $^1/_{200}$ of the value of $\sigma_a(730) - \sigma_a(420)$. If such a change in conductivity $(1.1 \cdot 10^{-13} \Omega^{-1})$ indeed took place, it would be possible to determine the lifetime τ_q from the relation

$$\delta\sigma_0 = [\sigma_a(730) - \sigma_a(420)] \frac{\Delta t}{\tau_q},$$

or

$$\tau_q = \frac{\sigma_a(730) - \sigma_a(420)}{\delta\sigma_0} \Delta t = 70 \cdot 200 = 14{,}000 \text{ hr.} \qquad (3.3)$$

Since $\delta\sigma_0$ is the upper limit of the possible change of σ_a during the 70 hours period, the value of τ_q obtained from (3.3) represents the lower limit of the lifetime. In other words, the carrier lifetime in a long-life trap (s-center) is over 14,000 hours or 1.6 years.

Prolonged observation of relaxation curves (over 100—120 hours) was carried out for a number of samples. In all cases, analogous results were obtained (see Figure 5). Following a fairly extended drop, a constant time-independent conductivity is observed during some tens of hours, which proves that the anomalous photoconductivity is time-independent and that the lifetime of a carrier in an s-center is quite large (over 1.6 year).

Time-independence of anomalous photoconductivity is also observed in relaxation curves recorded following blue-light illumination. Figure 20 shows a part of the conductivity relaxation curve after illumination with $\lambda = 420\,nm$. Unlike the relaxation curve observed after illumination at $\lambda = 730\,nm$ (when the conductivity drops in the dark), switching off the light in this case makes the conductivity first increase, approaching the $\sigma_a(420)$ level, at which it sterilizes and remains time-independent. The increase of conductivity observed after switching off the blue light will be explained in Chapter 4. Here we shall only note one basic fact: the presence on the relaxation curve of a section with the conductivity remaining constant for many tens of hours. This confirms the time-independence of $\sigma_a(\lambda)$.

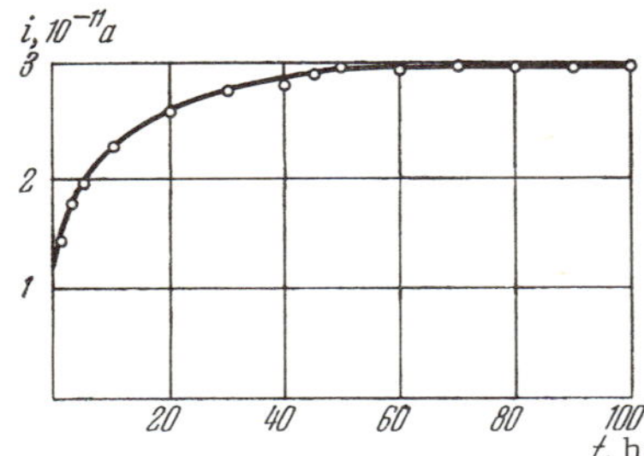

FIGURE 20. The relaxation of the photo-current in one of the anomalously photo-conductive selenium samples following illumination at 420 nm

All the existing experimental data on conductivity relaxation curves support the assumption that the carrier life-time in a long-life trap τ_q is over 1.6 year.

3. DEPENDENCE OF THE CROSS-SECTION FOR THE EXCITATION OF A CARRIER FROM A LONG-LIFE TRAP ON PHOTON ENERGY

In the phenomenological theory, the s-centers are fully characterized by the two parameters S and β'. The physical properties of these centers therefore can be inferred from the variation of these parameters under the influence of various factors. We shall consider the change in the cross-section S with the photon energy $\varepsilon = h\nu$.

The phenomenological theory suggests that if anomalously photoconductive selenium is illuminated with light of a wavelength different from the wavelength of the previous illumination, conductivity relaxation will be observed regardless of the time lag between the two illuminations. This relaxation can be expressed as

$$\sigma_a(\lambda) - \sigma_a = [\sigma_a(\lambda) - \sigma^{(0)}]\, e^{-SLt},$$

where $\sigma_a(\lambda)$ is the steady-state anomalous photoconductivity for the given wavelength λ, σ_a is the current value in the process of conductivity relaxation, $\sigma^{(0)}$ is the value of sample conductivity at the time of switching on the illumination λ, L is the light intensity, t is time, and S is the cross-section for the excitation of a carrier from a long-life trap by a photon. It follows from the above expression that the cross-section can be determined graphically using the dependence

$$\ln[\sigma_a(\lambda) - \sigma_a] = \ln[\sigma_a(\lambda) - \sigma^{(0)}] - SLt,$$

which defines a straight line given with a slope SL. This method requires the construction of the entire relaxation curve and the separation of the nonphotoactive component from the experimentally determined dependence $\sigma = \sigma(t)$. The relaxation time $\tau(\lambda)$ and the cross-section S can be determined from the steady-state value of anomalous photoconductivity and the initial section of the relaxation curve, namely

$$\tau(\lambda) = \frac{\sigma_a(\lambda) - \sigma^{(0)}}{\left(\dfrac{d\sigma_a}{dt}\right)_{t \to 0}} ; \qquad S = \frac{1}{\tau(\lambda) \cdot L}.$$

The determination of S by both the first and the second method assumes that the relaxation curve $\sigma = \sigma(t)$ in fact represents the relaxation of anomalous photoconductivity and that this dependence is not distorted by the presence of various nonphotoactive regions in the sample. Therefore, prior to the experiments, some control measurements had to be carried out. The underlying idea was that if $\sigma_a(t)$ indeed represented the relaxation of anomalous photoconductivity, the quantity

$$\frac{\sigma_a(\lambda) - \sigma^{(0)}}{\left(\dfrac{d\sigma_a}{dt}\right)_{t \to 0}}$$

should have the same value, irrespective of the initial conductivity $\sigma^{(0)}$ from which relaxation begins following illumination at the given wavelength. If, however, the observed shape of the $\sigma_a(t)$ curve is determined jointly by the variation of anomalous photoconductivity and other, secondary factors, the measured valve of $\left(\dfrac{d\sigma_a}{dt}\right)_{t\to 0}$ will in general deviate from $\sigma_a(\lambda) - \sigma^{(0)}$, i.e., the value of $\tau(\lambda)$ determined from $\left(\dfrac{d\sigma_a}{dt}\right)_{t\to 0}$ and $\sigma_a(\lambda) - \sigma^{(0)}$ will depend on the initial conductivity $\sigma^{(0)}$ at the time when the illumination is switched on.

All the values of $\tau(\lambda)$ presented below were determined by the second method. The value of $\sigma^{(0)}$ can be varied by the following two methods.

1. By varying the wavelength of the monochromatic light until a steady state is established in the sample: the steady-state conductivity is different for different wavelengths.

2. By illuminating the sample with monochromatic light of constant wavelength but varying duration (different irradiation doses).

These two methods of setting up different values of $\sigma^{(0)}$ give the same σ_a but differ in the nonanomalous component and its distribution across the selenium layer.

Control measurements were carried out by both methods of varying $\sigma^{(0)}$. The values of $\tau(\lambda)$ calculated from the two series of measurement turned out to be independent of the initial sample conductivity $\sigma^{(0)}$, and identical within the limits of experimental error. The contribution of nonphotoactive sections of the sample to the determination of $\tau(\lambda)$ was immaterial and was thus ignored.

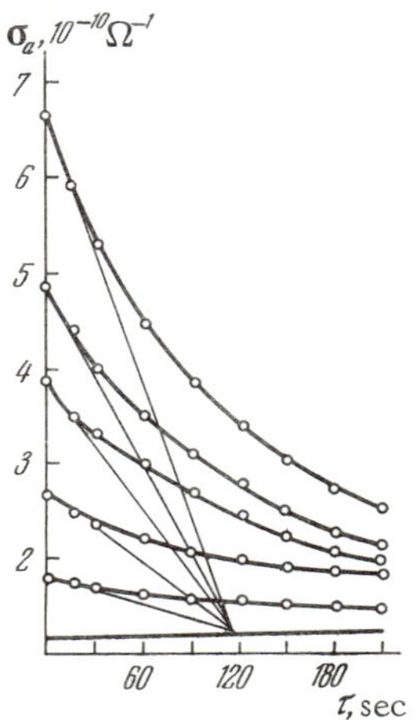

FIGURE 21. Determination of $\tau(\lambda)$

Figure 21 may serve as an illustration of the correctness of this statement. It shows relaxation curves for $\lambda = 420\,\mathrm{nm}$ with different initial values of the anomalous conductivity of the sample. The tangents to all the relaxation curves intersect approximately at one point, where the conductivity is equal to the steady-state value of anomalous photoconductivity for $\lambda = 420\,\mathrm{nm}$.

To investigate the dependence $S = S(\varepsilon)$, where $\varepsilon = h\nu$ is the energy of the incident photons, the sample was placed in a special cryostat /24/. A monochromatic light beam from a 3MR-3 monochromator was directed at the sample. The light intensity was determined by a calibrated bolometer.

For each wavelength investigated, the steady-state anomalous photoconductivity $\sigma_a(\lambda)$ and the value of $\left(\dfrac{d\sigma_a}{dt}\right)_{t\to0}$ were determined.

From these data, $\tau(\lambda)$ and then also $S = \dfrac{1}{\tau(\lambda)\cdot L}$ were calculated.

The determination of the spectral distribution of S requires careful measurements and constitutes a very labor-consuming task. The number of such investigations is therefore comparatively small /24, 25/. Note, however, that the shape of $S(\varepsilon)$ as presented in paper /25/ is not accurate: a correction was introduced in /25/ in the experimental value of $\left(\dfrac{d\sigma_a}{dt}\right)_{t\to0}$ based on the assumption that the long-life traps are uniformly distributed in the layer. Yet, as shown in Sec. 1 of the present chapter, the long-life traps are located in a layer close to the surface, and the use of the correction has introduced an error in the determination of S and has changed the shape of the function $S(\varepsilon)$. Paper /24/ is free from this inaccuracy and the investigation of the function $S(\varepsilon)$ is carried out more carefully.

Figure 22 plots the dependence of $\log S$ on the photon energy ε, based on data from /24/. Curve I corresponds to sample No. 3VK. The cross-section S varies according to a nearly exponential law. It is remarkable that such a dependence holds right up to photon energies of 2.9 eV, i. e., to energies much greater than the width of the energy gap in amorphous selenium.

Another feature of the curve is the presence of a maximum at a photon energy $\varepsilon = 1.8\,\mathrm{eV}$.

The dependence of S on ε (Figure 22) is quite stable. Thus, repeated measurements after an interval of 10 days gave the same values of S as that shown in curve I. During the interval, the sample had been heated twice to room temperature and subjected to continuous measurements. Substantial and irreversible changes took place in the sample after having been kept for 2 months at 40°C. The color of the layer changed markedly, the values of $\sigma_a(\lambda)$ and S

also changed, but the photoconductivity remained anomalous. The dependence of S on ε for the new state of the sample is shown on the same Figure 22 (curve II). The general shape of the curve remains the same. The position of the maximum does not change. Only the slope of the curve in the energy range $\varepsilon = 2.4-2.9\,\mathrm{eV}$ changes somewhat. The meaning of this will be clarified in the following section.

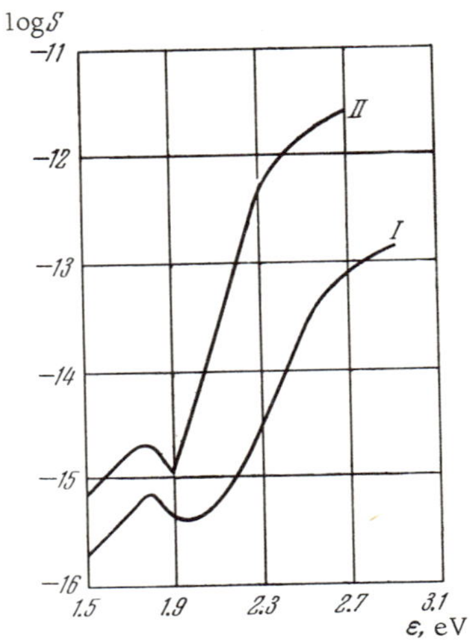

FIGURE 22.   Cross-section S vs. photon energy $\varepsilon = h\nu$

4. THE SHAPE OF THE POTENTIAL BARRIER SURROUNDING A LONG-LIFE TRAP

Some conclusions concerning the potential function of an s-center can be drawn from its basic properties postulated in Chapter 2. Since an s-center stores a captured electron for an indefinitely long time, its potential function obviously must constitute a deep potential well. At the same time, equilibrium carriers cannot penetrate into an s-center, certainly not in the temperature range in which anomalous photoconductivity exists. The probability of such an event is so small that it is neglected. An s-center thus constitutes a potential barrier for equilibrium electrons in selenium. These properties compel us to assume that an s-center is in fact a deep potential well surrounded by a potential barrier.

A more refined picture of the potential function for the interaction of an s-center with an electron can be obtained from experimental data on the dependence of the cross-section S on the photon energy ε, plotted in Figure 22. From these data the individual parameters of the potential function can be calculated. Let us first reiterate that the almost exponential relation between S and ε extends right up to $\varepsilon \simeq 3\,eV$. This means that the binding energy of an electron in an s-center is approximately 3 eV. Since the gap width in amorphous selenium is 2.3—2.5 eV /12, 26/, it is evident that the energy required to liberate a carrier from a long-life trap is much greater than the gap width. In this respect, there is a significant difference between the s-centers, on the one hand, and the impurity and recombination centers, on the other. Whereas the local energy levels corresponding to impurity and recombination centers fall within the energy gap, and the energy required for charge transfer from a localized center into the corresponding allowed band does not exceed E_g and is usually much smaller, for an s-center this energy is greater than E_g and, consequently, the localized energy level of an electron in an s-center may lie outside the selenium gap.

This suggests that the nature of s-centers is essentially different from the nature of recombination and trapping centers. Additional data supporting this conclusion will be presented later.

The fact that the binding energy of a carrier in a localized level in a long-life trap considerably exceeds the gap width in amorphous selenium confirms the previous conclusion that the potential function for the interaction of a carrier with a long-life trap cannot be represented by a simple potential well. This conclusion is also confirmed by the rate of change of the cross-section S with the energy ε.

Indeed, for a potential well of any shape, if the energy E acquired by the particle is less than its binding energy E_b the probability w of the particle escaping from the potential well can be expressed as

$$w = Ae^{-\dfrac{E_b - E}{kT}}, \qquad (3.4)$$

where k is the Boltzmann constant.

According to (3.4), the ratio of escape probabilities of a carrier from a well for two energies E_1 and E_2 is

$$\frac{w_1}{w_2} = e^{\dfrac{E_1 - E_2}{kT}}.$$

But the probability of a carrier escaping from a long-life trap by absorbing a photon is expressed through the cross-section S. Therefore, if the potential function for the interaction between a carrier and a long-life trap could be represented by a simple potential well, the following relation would apply:

$$\ln \frac{S_1}{S_2} = \frac{\varepsilon_1 - \varepsilon_2}{kT},\tag{3.5}$$

i. e., the rate of change of S as a function of photon energy should be such that the cross-section would change by a factor e when the photon energy changed by kT.

Yet it follows from the experimental data (curves I and II of Figure 22) that this variation is much slower. The dependence of S on ε was investigated at $T = 120°K$. At this temperature, $kT = 10^{-2}$ eV, whereas S changes by a factor e when the photon energy changes by 0.1 eV, i. e., roughly an order of magnitude slower than the prediction of relation (3.5). The fact that a change in the cross-section S by a factor of e requires a change of $10\,kT$ in energy and that the binding energy of a carrier in a trap exceeds the gap width confirms the model of a long-life trap as a potential well surrounded by a potential barrier. A carrier can leave the trap by tunneling across the barrier, not by passing over its top.

Such a potential function of a long-life trap provides a natural interpretation of the maximum on the log S curve (Figure 22). This maximum can arise as a result of resonant penetration of carriers across the potential barrier to a definite allowed energy level of the current carriers in the medium surrounding the long-life trap.

Since the cross-section S characterizes the probability of tunneling across a potential barrier, and we know the variation of S over a considerable energy interval $(1.5-2.9\,\text{eV})$, comparable to the height of the potential barrier, it appears to be possible to draw some conclusions about the shape of the potential barrier from the dependence $S\,(\varepsilon)$.

To be specific, we shall assume that a long-life trap captures an electron. Having absorbed a photon, the electron makes a transition to a higher energy state; although it stays only a short time in this state, its chances to pass across the potential barrier increase markedly as a result of the increase in its energy.

Let μ be the probability that an electron situated in a long-life trap absorbs a photon, v the electron velocity, and Δt the lifetime of the energy state $E_i + h\nu = E$, where E_i is the ground-state energy of an electron in a long-life trap. We shall denote by D_E the transmission of the barrier for an electron of energy E, and by a_0 the width of the potential well. We have

$$S = \mu\Delta t \cdot \frac{v}{a_0}\, D_E = C \cdot D_E,\tag{3.6}$$

where

$$C = \mu \Delta t \cdot \frac{v}{a_0} .$$

The transmission of the barrier depends exponentially on the energy E,

$$D_E = D_0 \exp\left[-\frac{4\pi}{h} \sqrt{2m} \int_{r_1}^{r_2} (U - E)^{1/2} \, dr\right], \qquad (3.7)$$

where U is the potential function, $\Delta r = r_2 - r_1$ is the width of the barrier at the height of the energy level E, D_0 is a factor which depends on the properties of the medium surrounding the barrier. For a barrier in air, $D_0 \simeq 1$. The other factors in (3.6) depend little on the energy, so that the observed variation of the cross-section S is mainly due to the variation of D_E as a function of ε. The barrier transmission D_E at a given energy level E is mainly determined by the width of the barrier. Therefore, the dependence of S on ε must contain information about the shape of the potential barrier, i.e., the dependence of its width on energy. In the anomalously photoconductive selenium sample investigation (No. 3VK), S increases by almost three orders of magnitude when ε is raised from 2.0 to 2.5 eV.

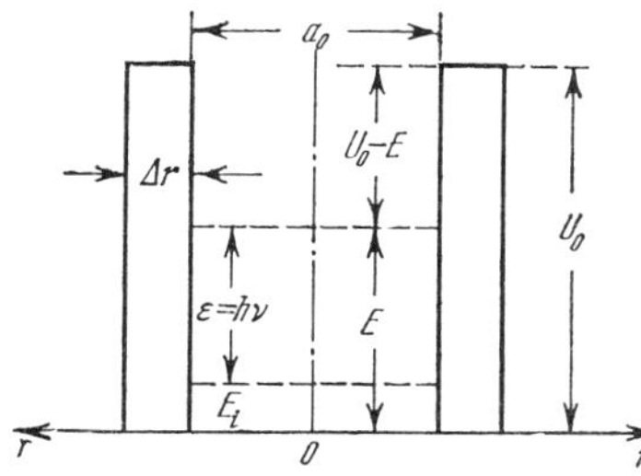

FIGURE 23. A potential function in the form of a potential well with a barrier U_0 of constant width Δr. E_i is the ground-state energy

When the photon energy is further increased, the rate of change of S noticeably decreases. Let us establish whether such a dependence $S(\varepsilon)$ can be derived from the model of a potential well surrounded by a rectangular barrier whose width is independent of the carrier energy (Figure 23). With such a barrier, the integrand $(U-E)^{1/2}$ does not depend on r and we have

$$D_E = \exp\left[-\frac{4\pi}{h} \sqrt{2m} \, \Delta r \, (U_0 - E)^{1/2}\right], \qquad (3.8)$$

where Δr is the barrier width, U_0 the barrier height. It follows from (3.6) and (3.8) that

$$\ln S = A - B \ (U - E)^{1/2}\Delta r,$$

where

$$A = \ln\left(\mu\,\Delta t\,\frac{v}{a_0}\right), \qquad (3.9)$$

$$B = \frac{4\pi}{h}\sqrt{2m}. \qquad (3.10)$$

Since

$$U_0 - E = E_b - \varepsilon,$$

where E_b is the binding energy (in Figure 23, it is the distance from the level at which the electron is located to the top of the potential barrier), $\varepsilon = hv$ is the photon energy, we have

$$\ln S = A - B\Delta r(E_b - \varepsilon)^{1/2}. \qquad (3.11)$$

Since the variation of S is mainly due to the second term in (3.11), A can be taken independent of ε, and then

$$\frac{d\ln S}{d\varepsilon} \simeq \frac{B}{2}\frac{\Delta r}{(E_b - \varepsilon)^{1/2}} . \qquad (3.12)$$

Expression (3.12) shows that with a rectangular barrier (Δr does not depend on ε) the derivative $\frac{d}{d\varepsilon}\ln S$ must increase with the energy ε, whereas the experimental data (Figure 22) show that the derivative $\frac{d}{d\varepsilon}\ln S$ actually decreases with the increasing ε.

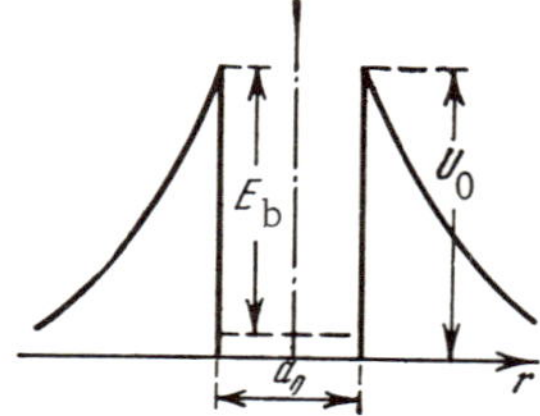

FIGURE 24. The potential function of a long-life trap in the form of a potential well surrounded by a Coulomb potential barrier

This leads to the conclusion that the potential barrier surrounding the trap must be different from that shown in Figure 23. Although this conclusion was drawn on the assumption that A does not depend on ε, nonetheless it remains valid in a more rigorous treatment, which takes into account the variation of A as a function of the energy.

Consequently, the barrier surrounding a long-life trap is not rectangular and the barrier width cannot be independent of ε. In order that $\frac{d}{d\varepsilon}\ln S$ should decrease (and not increase) with increasing ε, it is necessary that Δr should decrease with increasing ε, i. e., the potential barrier surrounding a long-life trap should taper toward the top.

Further, in order that the derivative $\frac{d}{d\varepsilon}\ln S$ should remain finite for $\varepsilon = E_b$ (in Figure 23, $\ln S$ varies slowly at $\varepsilon = E_b$, and $\frac{d}{d\varepsilon}\ln S$ tends to zero), it is necessary that the barrier width should vanish at $\varepsilon = E_b$ (i. e., at the top of the potential barrier).

Thus, it follows from the properties of s-centers, and in particular from the dependence of the carrier excitation cross-section of the photon energy, that an s-center constitutes a potential well more than $3\,\mathrm{eV}$ deep surrounded by a potential barrier tapering to zero width at its top.

This kind of barrier is characteristic of any charged center. We have therefore tried to interpret this barrier as a Coulomb barrier (Figure 24). With a Coulomb barrier it proved possible to obtain the experimental dependence of S on ε according to (3.6) and (3.7) by a proper choice of the barrier parameters.

5. THE PARAMETERS OF THE POTENTIAL BARRIER SURROUNDING A LONG-LIFE TRAP

We shall thus assume that we are dealing with a Coulomb barrier (Figure 24), i. e., $U(r) \sim \frac{1}{r}$. Let us determine the barrier parameters making use of (3.6) and (3.7).

The transmission D of a Coulomb barrier for an electron which has absorbed a photon ε is expressed by the relation

$$\ln D = -\frac{4\pi}{h}\sqrt{2m}\,r_0 U_0^{1/2}\times$$

$$\times\left[\left(\frac{U_0}{E_i+\varepsilon}\right)^{1/2}\arctan\left(\frac{U_0-E_i-\varepsilon}{E_i+\varepsilon}\right)^{1/2}-\left(\frac{U_0-E_i-\varepsilon}{U_0}\right)^{1/2}\right]+\ln D_0, \quad (3.13)$$

and the energy derivative of $\ln D$ is expressed by

$$\frac{d}{d\varepsilon}\ln D = \frac{4\pi}{h}\sqrt{2m}\,\frac{r_0}{2}\left(\frac{1}{E_i+\varepsilon}\right)^{1/2}\times$$
$$\times\left[\left(\frac{U_0}{E_i+\varepsilon}\right)\arctan\left(\frac{U_0-E_i-\varepsilon}{E_i+\varepsilon}\right)^{1/2}+\left(\frac{U_0-E_i-\varepsilon}{E_i+\varepsilon}\right)^{1/2}\right],\qquad(3.14)$$

where E_i is the ground-state energy and $U_0 - E_i = E_b$ determines the binding energy of a carrier in a long-life trap.

It follows from (3.13) and (3.14) that the variation of D and $\frac{d}{d\varepsilon}\ln D$ with ε is determined by the three parameters E_i, U_0 and r_0. To find these parameters, we need three equations. As a first approximation, it was assumed that $\mu\Delta t\cdot\frac{v}{a_0}$ did not depend on the photon energy and that the variation of S with ε was fully determined by the variation of D with ε. The following relations are then valid:

$$\ln\frac{D_{\varepsilon=\varepsilon_1}}{D_{\varepsilon=\varepsilon_2}} = \ln\frac{S_{\varepsilon=\varepsilon_1}}{S_{\varepsilon=\varepsilon_2}}\,,$$
$$\frac{d}{d\varepsilon}\ln D_{\varepsilon=\varepsilon_1} = \frac{d}{d\varepsilon}\ln S_{\varepsilon=\varepsilon_1},\qquad(3.15)$$
$$\frac{d}{d\varepsilon}\ln D_{\varepsilon=\varepsilon_2} = \frac{d}{d\varepsilon}\ln S_{\varepsilon=\varepsilon_2},$$

where ε_1 and ε_2 are two values of the photon energy chosen from the range of values used in the experiment. In the calculation performed, ε_1 was taken as $2.2\,\text{eV}$, and ε_2 as $2.7\,\text{eV}$ for curve I and 2.0, 2.5 respectively for curve II (Figure 22). It turned out that $E_b \simeq U_0$, $E_i = U_0 - E_b \ll U_0$.

If the long-life trap and its surrounding medium selenium are in equilibrium, both must have the same chemical potential level. However, since the trap has a long life-time, the charge transfer processes between the trap and the surrounding medium proceed slowly and such an equilibrium state may not be reached. This problem has to be resolved in special experiments. If, however, an equilibrium is established between the trap and the medium, and if the material of the trap is a conductor (a metal), the ground state of the electron in the trap coincides with the Fermi level, i.e., $E_i = E_F$. In this case the condition $U_0 - E_b \ll U_0$ indicates that the zero level of an s-center is not far from the chemical potential level of selenium.

To improve further the accuracy of the potential function parameters of a long-life trap, the dependence of $(\mu\Delta t\,v/a_0)$ on ε was taken into account.

The dependence of the parameters μ, Δt and v was roughly assessed, since it introduces only a small correction in the graph of S vs. ε. The microscopic absorption coefficient of a photon by an s-center μ was taken proportional to $\lambda^3 \sim 1/\varepsilon^3$. The life-time of a carrier in an excited state Δt was estimated on the basis of the uncertainty relation, i.e., $\Delta t \sim 1/\varepsilon$. As regards v, it was assumed, in accordance with the virial theorem, that

$$\frac{1}{2}\, mv^2 = E_{\text{kin}} = -E = U_0 - E_i - \varepsilon = E_b - \varepsilon.$$

Therefore v was taken proportional to $(E_b - \varepsilon)^{1/2}$. Under these assumptions,* relation (3.6) is transformed into

$$S = C' \frac{(E_b - \varepsilon)^{1/2}}{\varepsilon^4} \cdot D, \qquad (3.16)$$

or

$$\ln S = \ln C' + \ln \frac{(E_b - \varepsilon)^{1/2}}{\varepsilon^4} -$$
$$- \frac{4\pi}{h} \sqrt{2m}\, r_0 U_0^{1/2} \left[\left(\frac{U_0}{E_i + \varepsilon}\right)^{1/2} \arctan \left(\frac{E_b - \varepsilon}{E_i + \varepsilon}\right)^{1/2} - \left(\frac{E_b - \varepsilon}{U_0}\right)^{1/2} \right], \qquad (3.17)$$

where C' is independent of ε. Consequently,

$$\ln \frac{S_{\varepsilon=\varepsilon_1}}{S_{\varepsilon=\varepsilon_2}} = \ln \frac{(E_b - \varepsilon_1)^{1/2}\, \varepsilon_2^4}{(E_b - \varepsilon_2)^{1/2}\, \varepsilon_1^4} -$$
$$- \frac{4\pi}{h} \sqrt{2m}\, r_0 U_0^{1/2} \left\{ \left[\left(\frac{U_0}{E_i + \varepsilon_1}\right)^{1/2} \arctan \left(\frac{E_b - \varepsilon_1}{E_i + \varepsilon_1}\right)^{1/2} - \left(\frac{E_b - \varepsilon_1}{U_0}\right)^{1/2} \right] -$$
$$- \left[\left(\frac{U_0}{E_i + \varepsilon_2}\right)^{1/2} \arctan \left(\frac{E_b - \varepsilon_2}{E_i + \varepsilon_2}\right)^{1/2} - \left(\frac{E_b - \varepsilon_2}{U_0}\right)^{1/2} \right] \right\}. \qquad (3.18)$$

Equation (3.18) contains a number of experimental parameters S_{ε_1}, S_{ε_2}, ε_1, ε_2 and also three unknown parameters U_0, E_b, and r_0 of the potential function of the s-centers. It is essentially one of the three equations necessary for the calculation of these parameters.

Supplementing (3.18) with two additional relations

$$\left(\frac{d \ln S}{d\varepsilon}\right)_{\varepsilon=\varepsilon_1} =$$
$$= \frac{4\pi}{h} \sqrt{2m}\, \frac{r_0}{2} \left(\frac{1}{E_i + \varepsilon_1}\right)^{1/2} \left[\frac{U_0}{E_i + \varepsilon_1} \arctan \left(\frac{E_b - \varepsilon_1}{E_i + \varepsilon_1}\right)^{1/2} +$$
$$+ \left(\frac{E_b - \varepsilon_1}{E_i + \varepsilon_1}\right)^{1/2} \right] + \frac{1}{2}\, \frac{1}{E_b - \varepsilon_1} - \frac{3}{\varepsilon_1}, \qquad (3.19)$$

---

* Note that taking into account the dependence of μ and Δt on energy $\left(\mu \Delta t \sim \frac{1}{\varepsilon^4}\right)$ does not in practice affect the calculated value of the potential barrier parameters. Introducing the dependence of v on ε slightly changes the dependence of $\ln S$ on ε only for values of ε close to $U_0 - E_i$.

$$\left(\frac{d}{d\varepsilon}\ln S\right)_{\varepsilon=\varepsilon_2} =$$

$$= \frac{4\pi}{h}\sqrt{2m}\;\frac{r_0}{2}\left(\frac{1}{E_i+\varepsilon_2}\right)^{1/2}\left[\frac{U_0}{E_i+\varepsilon_2}\arctan\left(\frac{E_b-\varepsilon_2}{E_i+\varepsilon_2}\right)^{1/2} - \right.$$

$$\left. -\left(\frac{E_b-\varepsilon_2}{E_i+\varepsilon_2}\right)^{1/2}\right] + \frac{1}{2}\frac{1}{E_b-\varepsilon_2} - \frac{3}{\varepsilon_2}, \qquad (3.20)$$

we obtain three equations for the parameters U_0, E_i and r_0 which determine the probability of photoionization of a carrier in a long-life trap.

$\frac{d}{d\varepsilon}(\log S)_{\varepsilon=\varepsilon_1}$, $\frac{d}{d\varepsilon}(\ln S)_{\varepsilon=\varepsilon_2}$ and $\log\frac{S_{\varepsilon=\varepsilon_1}}{S_{\varepsilon=\varepsilon_2}}$ were determined from curves I and II in Figure 22. The calculated values of the parameters U_0, E_b and r_0 are as follows:

1. Sample No. 3VK before heating at 40°C (curve I, Figure 22):

$$U_0 = 3.5 \pm 0.4\ \text{eV},$$
$$E_b = 2.97 \pm 0.02\ \text{eV}, \qquad (3.21)$$
$$r_0 = 37.5 \pm 6.6\ \text{Å}.$$

2. Sample No. 3VK after aging at 40°C (curve II, Figure 22):

$$U_0 = 3.0 \pm 0.4\ \text{eV},$$
$$E_b = 2.82 \pm 0.02\ \text{eV}, \qquad (3.22)$$
$$r_0 = 33.8 \pm 7\ \text{Å}.$$

The errors in the parameters U_0, E_b and r_0 were determined so that the $\log S$ curve did not exceed the limits of error of the experimental values of S.

It is seen from the above data that only E_b is determined with relatively high accuracy ($< 1.0\%$); U_0 and r_0 are determined with a considerable error, $\sim 10\%$ for U_0 and nearly 20% for r_0. Although the error in the determination of U_0 and r_0 is considerable, their large value, roughly $3.0\ \text{eV}$ for U_0 and $r_0 \sim 30\ \text{Å}$, is beyond doubt.

The extent to which these parameters, determined from the value of $\log S$ and $\frac{d}{d\varepsilon}\ln S$ at two points, fit the general $\log S = f(\varepsilon)$ curve is evident from Figure 25.

Curve I in the figure plots $\log S(\varepsilon)$ as calculated from (3.18)–(3.19) using the values of the parameters U_0, E_b and r_0 from (3.21) (sample before heating). The short bars represent the experimental values of log S, showing the corresponding limits of error.

Curve II in Figure 25 shows $\log S$ as a function of E calculated using the values of the parameters from (3.22) (sample No. 3VK after heating).

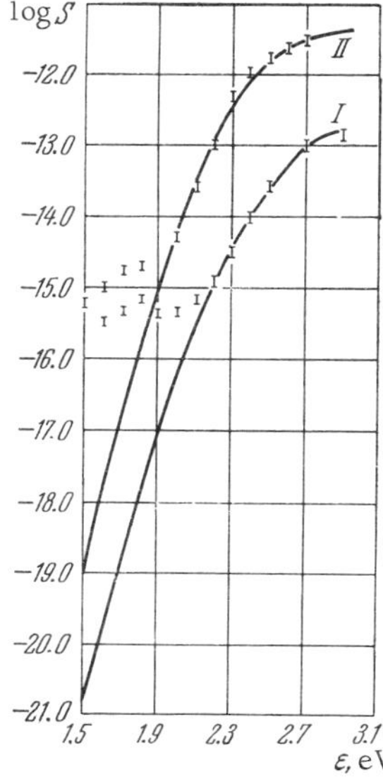

FIGURE 25. Calculated dependence of log s on ε. The vertical bars present the experimental data.

The curves show that the form assumed for the potential function of a long-life trap, i.e., a potential well surrounded by a Coulomb barrier with parameters (3.21) and (3.22), fits quite satisfactorily the experimental dependence log $S = f(\varepsilon)$.

The discrepancies between the calculated curves and the experimental data in the energy range 1.5–2.0 eV are due, as mentioned before, to the resonant penetration of electrons through the barrier, which is ignored in the calculation based on (3.18)–(3.19).

The data obtained on the potential function parameters of a long-life trap are of interest from two points of view. First, they give an idea of the properties of long-life traps which may serve as a basis for the construction of an appropriate physical model, and, second, they characterize the changes which take place when anomalously photoconductive selenium is markedly heated (at 40°C noticeable crystallization of an amorphous selenium begins in selenium layers) and which possibly also take place (although more slowly) at lower temperatures.

Among the parameters of s-centers in (3.21) and (3.22), the width of the potential well associated with an s-center $a_0 = 2r_0$ is of special interest. This quantity $a_0 = 75.6\pm18$ Å exceeds many times the mean interatomic distance in selenium. An s-center is thus not a center of atomic type. It is certain that this center is a complex formation with linear dimensions of a few tens of angstroms, consisting apparently of a large number of atoms. The size of this formation is close to the size of a colloidal particle.

The values of the potential barrier $U_0 \simeq 3\,\mathrm{eV}$ and of the binding energy $E_b = 2.9\,\mathrm{eV}$ are also astonishing. These values considerably exceed the width in selenium. This provides further evidence in favor of the complex nature of s-centers.

Prolonged heating of anomalously photoconductive selenium changes these parameters. Thus, the value $a_0 = 75.6 \pm 18\,\mathrm{Å}$ determined in the sample before heating decreased after heating to $a_0 = 67.6 \pm 14\,\mathrm{Å}$. The height of the potential barrier in the sample before heating was $U_0 = 3.5 \pm 0.4\,\mathrm{eV}$, and after heating it was $U_0 = 3.0 \pm 0.4\,\mathrm{eV}$. These changes are within the limits of measurement errors. The change in E_b is more significant. Although this change is outside the limits of experimental error, its magnitude is nevertheless negligible.

Consequently, the changes in an s-center are insignificant. This is in agreement with the fact that the shape of the $\log S\,(\varepsilon)$ curve does not change considerably by heating. The absolute value of S changes much more markedly. This is apparently associated not with the changes in the parameters of the long-life traps but with a change in the conditions characterizing the state of electrons in the selenium film.

It is of interest to clarify to what extent the potential function parameters of s-centers determined for sample No. 3VK are characteristic of the s-centers of mercury-activated amorphous selenium films. For this purpose, the potential function parameters of s-centers in other samples were determined by the above method. The results are summarized in Table 2.

TABLE 2

No. of sample	U_0, eV	r_0, Å	E_b, eV
3VK	3.5 ± 0.4	37.5 ± 6.6	2.97 ± 0.02
3VK after heating	3.0 ± 0.4	33.8 ± 7.1	2.82 ± 0.02
6VK	3.5 ± 0.4	37.5 ± 6.6	2.97 ± 0.02
7VK	3.5 ± 0.4	51 ± 7	2.96 ± 0.02
8VK	3.5 ± 0.4	44 ± 6	2.93 ± 0.02

It is seen from the data in Table 2 that the parameters U_0 and r_0 in all these samples are close to each other and their differences are within the limits of measurement error. As regards E_b, although the change in it is outside the limits of experimental error, the absolute value of this change is negligible.

Thus, we can conclude that in all the selenium samples investigated the potential function parameters of s-centers are fairly constant.

It is natural that the parameters of the potential function of long-life traps should be reflected in the model of an s-center developed in Sec. 6 below.

6. MODEL OF A LONG-LIFE TRAP AND ITS IMPLICATIONS

The data of Sec. 5 served as a basis for the construction of the following model of a long-life trap.

A long-life trap is a macroscopic formation which contains a large number of atoms (a few thousands). The trap can be treated as a subcolloidal particle. The potential function of a long-life trap is a deep spherical potential well 70–80 Å wide, surrounded by a Coulomb potential barrier. The barrier height is approximately 3 eV. The carriers (electrons) in a trap are degenerate, i. e., the trap is metallic (possibly mercury) or semiconducting with a high conductivity (possibly mercury selenide). Such a model of a long-life trap will subsequently be called the s u b c o l l o i d a l m o d e l.

Let us examine what properties of anomalous photoconductivity and of the long-life traps can be explained by the subcolloidal model.

a) The lifetime of a carrier in a long-life trap

In the phenomenological theory it was assumed that a carrier captured by a long-life trap is stored for an indefinitely long time, and can be liberated from the trap only by absorbing a photon. Any particular model of the trap must possess this property and we shall first clarify whether the subcolloidal model meets this requirement. The parameters of the trap (the width of the well and the height of the potential barrier) were determined from the dependence on photon energy of the cross-section for carrier excitation from the trap. The lifetime τ_q did not enter into this consideration. Therefore, the lifetime of a carrier in a well with such parameters is not a priori known. Does it exceed the experimental lower limit of $1\tfrac{1}{2}$ years or not?

Since the energy level of an electron in the potential well lies above the zero level of the potential barrier (0.16 eV in sample No. 3VK before heating and 0.18 eV after heating), there is some probability λ_i that the electron will leak through the barrier,

$$\lambda_i = \frac{v_0}{a_0} D_0, \qquad (3.23)$$

where v_0 is the electron velocity in the ground state inside the potential well, a_0 is the width of the well, D_0 the transmission of the barrier for an electron in the ground state. The lifetime τ_q is determined by the relation

$$\tau_q = \frac{a_0}{v_0}\, D_0^{-1}. \qquad (3.24)$$

D_0 is determined from (3.13) with the assumption $\varepsilon = 0$; v_0 is found from the relation

$$\frac{1}{2}\, m v_0^2 = E_i.$$

Computing v_0 and D_0 and substituting in (3.24) we find τ_q. Calculations were carried out for the two sets of values of the long-life trap parameters presented in the previous section. The lifetime of a carrier in the long-life traps in the sample was found to be $\sim 5.5 \cdot 10^{96}$ sec before heating and $\sim 2.8 \cdot 10^{93}$ sec after heating.

This value of τ_q is orders of magnitude greater than 1.6 years. Consequently, traps with the above parameters are indeed long-lived. An electron captured in such a trap will be stored for a long time (more than 10^{85} years), practically infinitely long. If selenium indeed contains such long-life traps, it is naturally impossible to observe any decrease of photoconductivity even after many days in the dark.

The truly long lifetime of an electron in a trap having the above parameters, i. e., the identification of the trap as a long-life trap, can be considered the first achievement of the subcolloidal model.

b) Temperature dependence of the lifetime of
a carrier in a long-life trap

It has already been mentioned earlier that in amorphous selenium samples treated with mercury vapor, the anomalous character of photoconductivity is observed only at quite low temperature. At room temperature, the photoconductivity is normal. This problem will be dealt with in detail in Chapter 4. Here, we shall only mention that one might think that a possible reason for the absence of anomalous photoconductivity at room temperature is the fact that the lifetime of an electron in a trap decreases with rising temperature to an extent that the s-centers, which at low temperature constitute long-life traps, cease to act as such at room temperature or near it.

Since the potential function parameters of an s-center are known, one can calculate how τ_q will vary as a function of temperature, what will be its value at room temperature, and consequently clarify whether the s-center at room temperature remains a long-life trap or turns into an ordinary recombination center.

The above lifetime of an electron in the long-life traps of anomalously photoconductive selenium is calculated on the assumption that the trap is metallic and that the electron escapes from the trap by tunneling from the level $E_F = U_0 - E_b$. But the electron occupies this state at the absolute zero temperature. At higher temperatures, the electron has a finite probability w_E to be in a state with energy E higher than the Fermi level E_F. In the subcolloidal model this probability is expressed by the Fermi function

$$w_E = \frac{1}{e^{\frac{E-E_F}{kT}} + 1}. \tag{3.25}$$

Since at a finite T an electron can be in states of different energy $E > E_F$, the probability λ_i that the electron will leak through the barrier is given

$$\lambda_i = \int w_E \frac{v}{a_0} D_E q(E)\, dE, \tag{3.26}$$

where D_E is the transmission of the barrier for an electron in a state with energy E, $q(E)$ is the concentration of electrons at this level. The integral in (3.26) is taken over all the states of the electron in the trap.

Although the probability w_E is small that the electron is in a state with $E - E_F$ considerably greater than kT, electrons will on the whole leak through the barrier precisely from such states. Since the Coulomb barrier becomes much broader at its base, the width of the barrier at small E will be very large in the model assumed (the transmission through a barrier of such width proceeds as mentioned above, with a probability $< 10^{-90}$); with increasing energy E, however, the barrier width decreases markedly and the probability of transmission increases very fast. Consequently, the fast variation of the barrier width with increasing E at small values of E leads, in the subcolloidal model, to a strong temperature dependence of the transmission probability of a trapped carrier. This problem was investigated in detail elsewhere /27/.

Figure 26 shows the temperature dependence of carrier lifetime in a long-life trap.

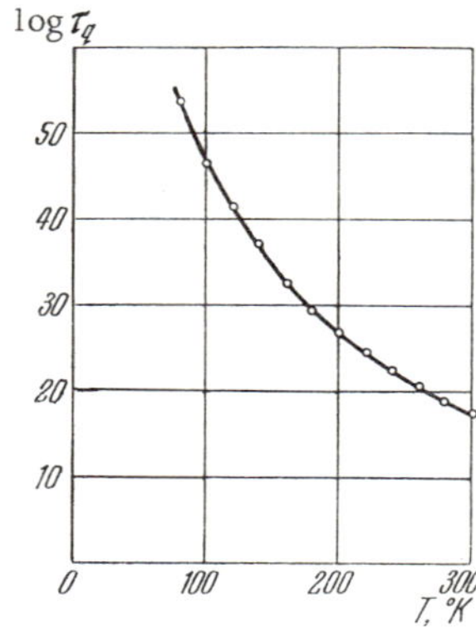

FIGURE 26. Calculated temperature dependence of τ_q (sample
No.3VK before heating)

As expected, τ_q varies very fast with temperature and drops
from approximately 10^{55} sec at 85°K to 10^{18} sec at room temperature. However, even this lifetime is very very long $(3 \cdot 10^{10}$ years$)$
and the s-center can accordingly be considered as a long-life trap.

Consequently, in the subcolloidal model, the s-center must be a
long-life trap both at low and at room temperatures, and the reason
for the disappearance of anomalous photoconductivity at temperatures $T > 200$°K is not associated with the reduction of the lifetime
τ_q with increasing temperatures.

c) Cross-section for photoexcitation of a carrier from a long-life trap

In Sec. 4 we considered the dependence of the cross-section S on
the photon energy. Here we would like to turn our attention to the
calculated value of S. In the short-wave region of the visible
spectrum, the value of S attains 10^{-12}–10^{-13} cm^2. This is very large.
Indeed, let us compare this value of the cross-section S for various
energies with the value of the absorption coefficient of selenium.
Note that the cross-section is in fact the absorption coefficient
referred to a single absorbing center. In Table 3 values of S are
compared with the atomic absorption coefficients $\mu = K/N_0$, where K
is the absorption coefficient of selenium, N_0 is the number of
selenium atoms in a unit volume.

Table 3 shows values of μ, S and their ratio for several wavelengths between 420 nm and 715 nm. These wavelengths were chosen
in such a way that S and μ could be compared both in the range of
the intrinsic absorption of selenium and outside it. The values of S
presented in this table are taken from the curve in Figure 22.

TABLE 3

λ, nm	715	640	580	500	200
S	$1.7\cdot10^{-15}$	$1.7\cdot10^{-15}$	$4\cdot10^{-14}$	10^{-12}	$3\cdot10^{-12}$
$\mu = K/N_0$	$2.65\cdot10^{-21}$	$4.45\cdot10^{-20}$	$7.45\cdot10^{-19}$	$5.3\cdot10^{-18}$	10^{-17}
S/μ	$6.4\cdot10^{5}$	$3.8\cdot10^{4}$	$5.4\cdot10^{4}$	$2.0\cdot10^{5}$	$3\cdot10^{5}$

According to the data of Table 3, the cross-section S exceeds μ by 4—5 orders of magnitude both inside and outside the absorption band. Note that the absorption coefficient of selenium in the intrinsic absorption region is large and reaches $3\cdot10^{5}$ cm^{-}. Such an absorption coefficient means that the radiation intensity is halved by passing through a layer $2\cdot10^{-6}$ cm thick. If the long-life traps were atomic centers and their atomic absorption coefficient had a value corresponding to a cross-section of $3\cdot10^{-12}$ cm^{2}, the radiation intensity would be halved by passing through a material layer 10^{-11} cm thick. It is evident that atomic centers cannot possess such a high absorption coefficient. In the subcolloidal model, in which a trap contains of the order of ten thousand atoms, the absorption of light by a trap must also be approximately 10^{4} larger than the absorption due to a single atom.

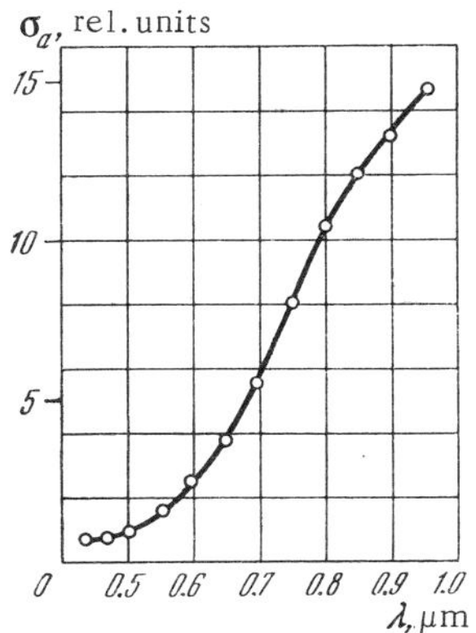

FIGURE 27. Spectral distribution of anomalous photoconductivity

d) Spectral distribution of anomalous photoconductivity

The spectral distribution of anomalous photoconductivity (Figure 27) has a nontrivial character. The steady-state photoconductivity under illumination is small in the intrinsic absorption range of amorphous selenium, it increases with wavelength across the absorption edge, and outside the absorption band it is much higher. The phenomenological theory of anomalous photoconductivity does not account for this feature.

From the phenomenological theory it only follows that the steady-state value of anomalous photoconductivity $\sigma_a (\lambda)$ is determined by the relation

$$\sigma_a (\lambda) = eu \frac{K(\lambda) \cdot \beta' (\lambda)}{S(\lambda)}, \qquad (3.27)$$

where u is the mobility of majority carriers in selenium, $K(\lambda)$ is the absorption coefficient for radiation of wavelength λ, $\beta'(\lambda)$ is the quantum yield of the process of electron capture by a long-life trap, $S(\lambda)$ is the cross-section for the excitation of an electron by a quantum. In the phenomenological theory (Chapter 2) the quantities β' and S are introduced formally and no conclusion is drawn as to their possible values, so that it is not possible to calculate the value of $\sigma_a (\lambda)$ from this theory. The fact that between $420-715$ nm, $\frac{d\sigma_a (\lambda)}{d\lambda} > 0$ was interpreted as indicating that $S(\lambda)$ decreases with increasing wavelength faster than the product $K(\lambda) \cdot \beta'(\lambda)$ does.

Since the parameters of the long-life traps have now been determined, it becomes possible to calculate $\frac{d\sigma_a (\lambda)}{d\lambda}$ from the curve $S(\varepsilon)$ both in the spectral interval investigated and outside it.

Let monochromatic light whose photon energy is greater than the gap width of selenium E_g fall on selenium containing long-life traps. Electrons will then appear in the conduction band with energies (relative to the bottom of the conduction band) in the interval from 0 to $\varepsilon - E_g$.

An electron having an excess energy $E - E_c$, where E_c is the energy of an electron at the bottom of the conduction band, will lose it over a pathlength l through interaction with the selenium. It is evident that the pathlength l increases with the energy $E - E_c$. We shall subsequently denote by l the free path of an electron of energy E. This free path will be determined by the lifetime of an electron in the free state ($\sim 10^{-7}$ sec).

A photoexcited electron thus spends part of the time in a state of energy E, and part of the time in a state of energy E_c. The probability of its penetration into a long-life trap is therefore made up of two parts: γ_E and γ_{E_C}. Let D_E and D_{E_C} be the transmission of the potential barrier for an electron in the states E and E_c, respectively, Q the concentration of long-life traps, and a_0 the width of the potential well.

We than have

$$\gamma = \gamma_E + \gamma_{E_C} = \pi a_0^2 \, Q \, (l_{E_C} D_{E_C} + l_E D_E), \qquad (3.28)$$

where

$$l_E D_E = \int_0^l D_{E'} dl, \qquad (3.29)$$

and E' is the energy of the carrier as it is being de-excited from E to E_C.

γ is the probability that an electron knocked out from a trap by absorbing a photon will be recaptured in a long-life trap. If γ is multiplied by $\beta \tau_0$ — quantum yield of the internal photoeffect times the lifetime of a free electron, we obtain the magnitude β' which constitutes the quantum yield of electron capture by a long-life trap:

$$\beta' = \beta \pi a^2 Q \, (l_E D_E + l_{E_C} D_{E_C}) \, \tau_0. \qquad (3.30)$$

Taking $\sigma_a(\lambda)$ from (3.27), $\dfrac{d\sigma_a(\lambda)}{d\lambda}$ can be found from the relation

$$\frac{d}{d\lambda} \sigma_a(\lambda) = - \, eu \, \frac{K(\lambda)\,\beta'(\lambda)}{S(\lambda)} \cdot \frac{hc}{\lambda^2} \left[\frac{d}{d\varepsilon} \ln K(\lambda) + \frac{d}{d\varepsilon} \ln \frac{\beta'(\lambda)}{S(\lambda)} \right]. \qquad (3.31)$$

Assuming that $\ln \dfrac{\mu \Delta t v}{a_0}$ is independent of energy, we may write

$$\frac{d}{d\varepsilon} \ln \frac{\beta'(\lambda)}{S(\lambda)} = \frac{d}{d\varepsilon} \ln \left(\frac{l_E D_{\varepsilon + \delta} + l_{E_C} D_{E_g + \delta}}{D_{E_i + \varepsilon}} \right). \qquad (3.32)$$

The quantity δ entering (3.32) determines the location of the electron energy levels in a long-life trap relative to those in selenium (Figure 28). This is the distance between the top of the valence band and the zero (ground) level of the trap.

Substituting (3.32) into (3.31), we obtain

$$\frac{d}{d\lambda} \sigma_a(\lambda) = - eu \frac{K(\lambda)\,\beta'(\lambda)}{S(\lambda)} \frac{hc}{\lambda^2} \left[\frac{d}{d\varepsilon} \ln K + \frac{d}{d\varepsilon} \ln \left(1 + \frac{l_E D_{\varepsilon + \delta}}{l_{E_C} D_{E_g + \delta}} \right) - \frac{d}{d\varepsilon} \ln D_{E_i + \varepsilon} \right]. \qquad (3.33)$$

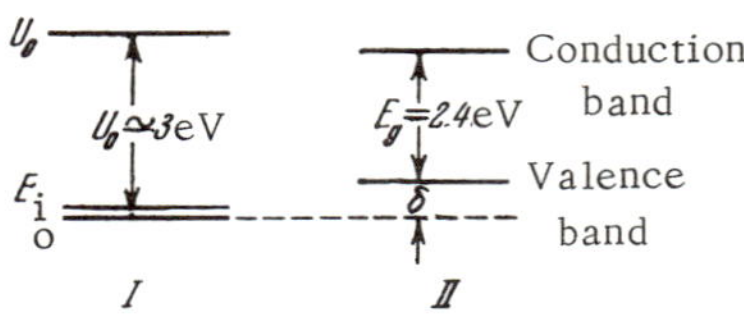

FIGURE 28. The determination of δ:
I shows the location of levels in a
long-life trap, II shows same in selenium.

Let us analyze expression (3.33). First we shall find the sign of $\dfrac{d\sigma_a(\lambda)}{d\varepsilon}$. This sign is determined by the relation of $\dfrac{d}{d\varepsilon}\ln K$ to $\dfrac{d}{d\varepsilon}\ln D$. Figure 29 shows curves characterizing the dependence of these magnitudes on energy. It is seen that $\left|\dfrac{d}{d\varepsilon}\ln D\right| > \left|\dfrac{d}{d\varepsilon}\ln K\right|$.

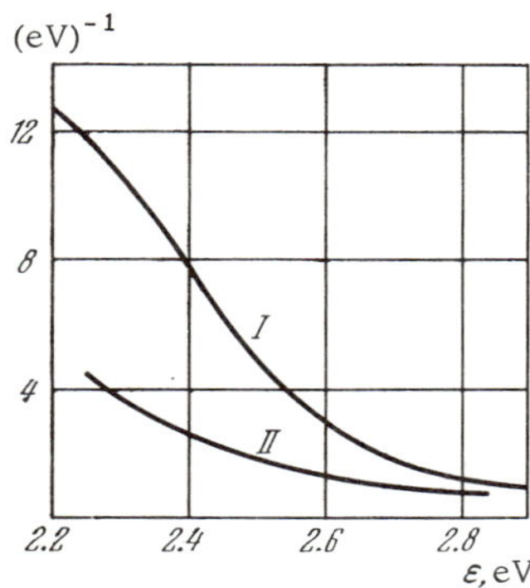

FIGURE 29. Illustrating the
spectral distribution of anomalous
photoconductivity: I shows
$\dfrac{d}{d\varepsilon}\ln D$ as a function of ε in
eV^{-1}, II shows $\dfrac{d}{d\varepsilon}\ln K$ as a function of ε in eV^{-1}.

When $\varepsilon = E_g$, i. e., for light whose photon energy is equal to the width of the energy gap in selenium, the middle term in (3.33) vanishes. In this case $\dfrac{d}{d\lambda}\sigma_a(\lambda)$ must be positive, i. e., when λ passes through the absorption edge of selenium, the anomalous photocon-ductivity, according to the subcolloidal model, should increase with λ, which is indeed so (whereas the ordinary photoconductivity de-creases as λ crosses the absorption edge).

When ε increases, the term

$$\frac{d}{d\varepsilon}\ln\left(1 + \frac{l_E D_{\varepsilon+\delta}}{l_{E_C} D_{E_g+\delta}}\right)$$

increases, whereas the difference $\frac{d}{d\varepsilon}\ln D_\varepsilon - \frac{d}{d\varepsilon}\ln K$ decreases; consequently the absolute value of $\frac{d\sigma_a(\lambda)}{d\lambda}$ will also decrease, which is also observed.

Thus, according to the subcolloidal model, $\frac{d\sigma_a(\lambda)}{d\lambda}$ is positive in the range of energies above $2.4\,\mathrm{eV}$ and decreases with increasing photon energy.

The variation of $\sigma_a(\lambda)$ for $\varepsilon < 2.4\,\mathrm{eV}$ will have a different character. At these energies, an electron falls into a trap through a resonance level. The quantum yield changes little with energy. The difference $\frac{d}{d\varepsilon}\ln K - \frac{d}{d\varepsilon}\ln D$ for $\varepsilon < 2.5\,\mathrm{eV}$ is also fairly constant. Therefore $\frac{d\sigma_a(\lambda)}{d\lambda}$ in the range $\varepsilon < 2.5\,\mathrm{eV}$ must be a positive quantity, little dependent on energy. Such a behavior (curve in Figure 6) is indeed observed. In this region $\frac{d\sigma_a(\lambda)}{d\lambda}$ is almost a linear function of ε.

e) The maximum of S on the $S(\varepsilon)$ curve

As mentioned before, the energy dependence of the excitation cross-section of an electron from a long-life trap, calculated using the long-life trap parameters of the subcolloidal model, agrees with the experimental data only in the energy range $2.2-2.9\,\mathrm{eV}$. At lower energies, however, the observed $S(\varepsilon)$ curve differs substantially from the calculated curve. The value of S actually increases (rather than decreases) with decreasing energy and attains a maximum at an energy $\varepsilon = 1.77\,\mathrm{eV}$. Such a maximum on the curve of the transmission probability of a particle through a barrier vs. the particle energy can arise in the case of resonance transmission, when the particle energy coincides with some allowed state on the other side of the barrier. A carrier transmitted through the potential barrier of a long-life trap enters the selenium medium and moves in it with the same total energy. It is thus evident that the resonance level of a long-life trap, which we shall denote by E_r, should coincide with the energy level of some allowed electron state in selenium.

To determine the shape of the resonance peak, the values of S calculated from the transmission probability through a potential barrier having the given parameters were subtracted from the experimental values of $S(\varepsilon)$ (short bars on curves I and II in Figure 25). Figure 30 plots the dependence $\Delta S = S_{exp} - S_{calc}$ for sample No. 3VK before heating, and Figure 31 plots this dependence for the same sample after heating. From the shape of the curves, it is possible to calculate the width of the resonance level ΔE_r. In this calculation, it must be borne in mind that the shape of the resonance peak, determined as the difference between the experimental values of $S(\varepsilon)$ and those calculated from the s-center parameters, depends little on the accuracy of the extrapolation of $S(\varepsilon)$ to energies $\varepsilon < \varepsilon_m$, where ε_m is the maximum energy of the resonance peak, whereas for $\varepsilon > \varepsilon_m$ this shape is highly sensitive to the way the theoretical curve $S(\varepsilon)$ has been constructed. Therefore, the width of the resonance must be determined from the part of the $\Delta S = S(\varepsilon)_{exp} - S(\varepsilon)_{calc}$ curve for $\varepsilon < \varepsilon_m$.

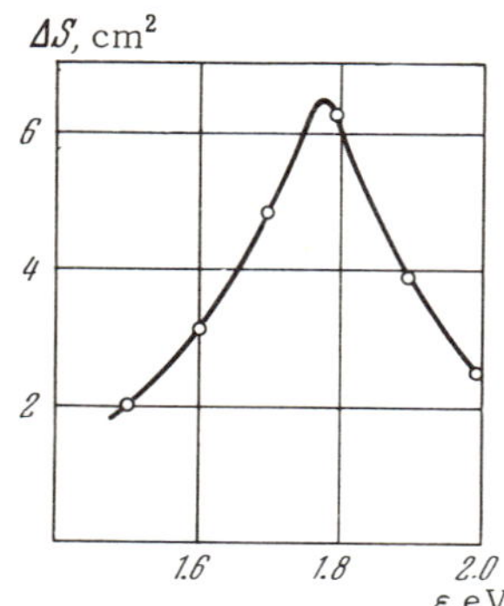

FIGURE 30. The shape of the resonance peak ΔS for sample No.3VK before heating

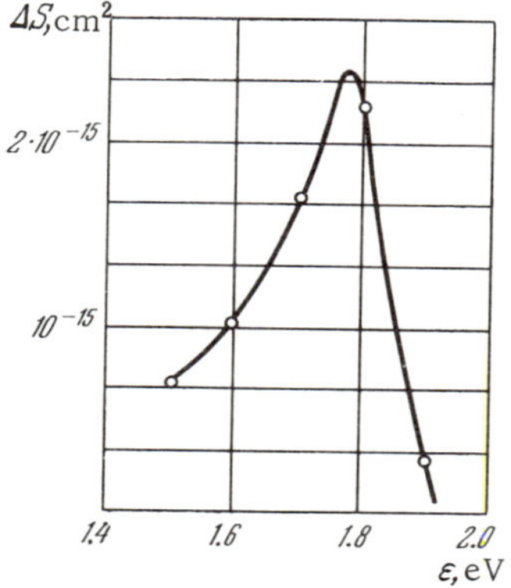

FIGURE 31. The shape of the resonance peak ΔS for sample No.3VK after heating

From the shape of the curves on Figures 30 and 31, it may be concluded that the halfwidth of the resonance peak is $0.1-0.15\,\text{eV}$ and that it varies little in the process of aging of the sample.

Summing up the foregoing, we note that the subcolloidal model, proposed on the basis of the experimental data on the dependence $S(\varepsilon)$, makes it possible to explain successfully a number of properties of long-life traps and anomalous photoconductivity of amorphous selenium. According to this model, the following points become clear:

a) The enormous lifetime of a carrier in a trap at low temperatures ($\tau_q \sim 10^{30}-10^{40}$ years).

b) The strong temperature dependence of the lifetime τ_q, and at the same time preservation of the properties of s-centers at room temperature: a carrier falling into a trap at room temperature will remain stored in dark for a long time (more than 10^{10} years).

c) The energy dependence of the cross-section S.

d) The relatively large value of the cross-section.

e) The spectral distribution of anomalous photoconductivity.

7. DEPENDENCE OF THE QUANTUM YIELD ON PHOTON ENERGY FOR THE CASE OF CARRIER EXCITATION INTO A LONG-LIFE TRAP

Let us now turn to the dependence on photon energy of β' — the quantum yield for carrier excitation into a long-life trap.

In the phenomenological theory the magnitude β' determines the level of anomalous photoconductivity

$$\sigma_a\,(\lambda) = \frac{eu_p K \beta'}{S} = e \cdot u_p \, \frac{K \beta \tau_0 \gamma Q}{S} \, . \tag{3.34}$$

Since the cross-section as a function of photon energy has been determined, and both the dependence of the absorption coefficient in selenium on the photon energy and the spectral distribution of anomalous photoconductivity are known, we can establish the energy dependence of the quantum yield

$$\beta'\,(\lambda) = \frac{\sigma_a\,(\lambda) \cdot S\,(\lambda)}{eu_p K\,(\lambda)} \, . \tag{3.35}$$

To eliminate the mobility u_p, we shall use the relative value of the quantum yield

$$\beta'_0\,(\lambda) = \frac{\beta'\,(\lambda)}{\beta'\,(\lambda_0)} = \frac{\sigma_a\,(\lambda) \cdot S\,(\lambda) \cdot K\,(\lambda_0)}{\sigma_a\,(\lambda_0) \cdot S\,(\lambda_0) \cdot K\,(\lambda)} \, , \tag{3.36}$$

where all the quantities subscripted 0 refer to illumination with light of wavelength $\lambda = 457\,\text{nm}$, while quantities without the subscript refer to the radiation of wavelength λ.

The quantity β'_0 defined by (3.36) can be represented also in the form

$$\beta'_0 = \frac{\beta\,(\lambda)\,\gamma\,(\lambda)\,\tau_0 Q}{\beta\,(\lambda_0)\,\gamma\,(\lambda_0)\,\tau_0 Q} = \frac{\beta\,(\lambda)\,\gamma\,(\lambda)}{\beta\,(\lambda_0)\,\gamma\,(\lambda_0)} \, . \tag{3.37}$$

If γ does not depend on λ, β_0 becomes the relative quantum yield in activated amorphous selenium films.   The starting point in the calculation of β_0' (according to (3.36)) was the fact established in /6/ that, in amorphous selenium films activated with mercury to the extent required for the appearance of anomalous photoconductivity, the absorption coefficient K in the 400—700 nm spectral range does not differ substantially from the absorption coefficient of pure amorphous selenium.   Therefore, the value of the absorption coefficient for amorphous selenium was substituted in (3.36).   Data on these coefficients are presented elsewhere /12, 28/. σ_a (ε) and S (ε) were determined experimentally for the particular sample investigated. The calculated values are shown graphically in Figure 32, where the abscissa represents the photon energy ε and the ordinate gives the logarithm of the relative quantum yield (the value of β' for $\varepsilon = 2.7$ eV is taken as 1).   The energy dependence of β_0' is highly interesting.

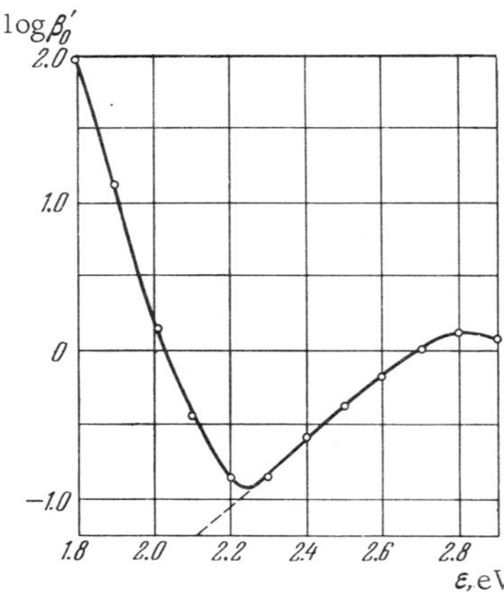

FIGURE 32. The logarithm of the relative quantum yield β_0' vs. photon energy ε

Note that according to (3.37) β_0' contains the relative quantum yield β (λ)/β (λ_0) as a factor.   It is quite possible that the second factor in this expression $\gamma(\lambda)/\gamma(\lambda_0)$ does not depend on the energy and that β' (ε) may turn out to be proportional to β (ε).   We shall clarify now whether such a proportionality holds.   For this purpose, we shall compare the curves β (ε) and β' (ε).

First, consider the variation of β' in the energy range $2.2 < \varepsilon < 2.9$ eV.   There are a number of investigations of the dependence of β on the photon energy ε in this energy range.   At the bottom of this energy range β is small and has a value of the order of 10^{-2}. It increases with ε and approaches unity as the photon energy approaches 2.9—3.0 eV.

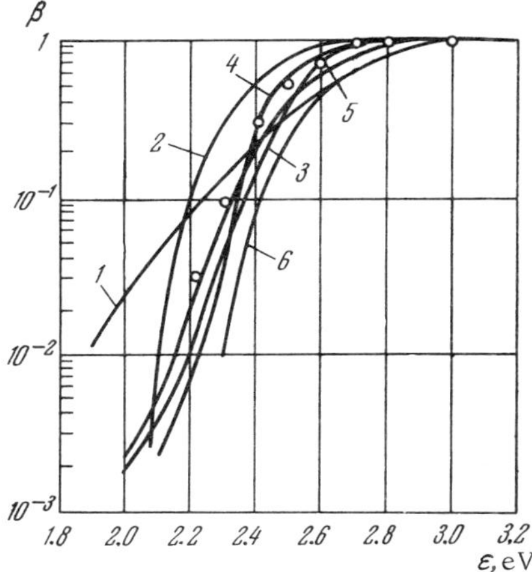

FIGURE 33. The quantum yield β of amorphous
selenium vs. the photon energy according to
data of various authors:

1) /29/; 2) /33/; 3) /31/; 4) /32/; 5) /11/;
6) /30/.

Figure 33 compares the functions β (ε) obtained by various investigators /11, 29—33/ with the dependence β' (ε) /80/. The continuous curves 1—6 in this figure represent the energy dependence
of the quantum yield β of amorphous selenium. The circles plot
the values of β' for various photon energies. The value of β' at
$\varepsilon = 3.0\,\text{eV}$ was taken as 1.

It is seen in Figure 33 that the points are fitted well by the curves
3, 4, 5. This means that β and β' have the same energy dependence.
Two conclusions follow from this fact.

1) In the spectral range corresponding to quantum energies ε from
2.2 to 2.9 eV, the probability γ that a minority carrier is captured
by an s-center does not depend on the photon energy.

2) The minority carriers excited in a mercury-activated amorphous selenium film by light in this spectral range reach the
s-centers through the conduction band. The value of β for mercury-
activated amorphous selenium films in the relevant spectral range
should be chosen the same as for pure selenium.

The situation is different in the energy range $\varepsilon < 2.2\,\text{eV}$. If an
electron excited in selenium as a result of the absorption of a
photon were to penetrate the s-centers only through the conduction
band, a drop of β_0' with decreasing energy ε should also be observed
for $\varepsilon < 2.3\,\text{eV}$. Such a drop in β_0' should be even more pronounced
than that observed in the energy interval 2.3—2.7 eV. In reality,

however, no drop in β_0' with decreasing ε takes place. For values of ε less than 2.2 eV, β_0' does not decrease with decreasing ε: it grows (Figure 32), increasing by almost three orders of magnitude as ε decreases from 2.2 to 1.8 eV. This variation of β_0' as a function of ε indicates that there are two, and not one, channels (two possibilities) for an electron to fall into a long-life trap. Accordingly, β_0' may be written as a sum of two quantities

$$\beta_0' = \beta_1' + \beta_2', \qquad (3.38)$$

where β_1' represents the quantum yield of electron capture by a long-life trap through the conduction band (first channel), and β_2' is the quantum yield of electron capture through the second channel. As to the nature of this second channel, it may be established from an analysis of the energy dependence of β_2'. To obtain this dependence on ε, β_1' must be separated from the sum β_0'. Such a separation can be achieved by assuming that in the energy range $\varepsilon = 2.3-2.7$ eV only one of the two channels is effective, and that in this energy range $\beta_0' = \beta_1'$. The function β_1' can then be obtained by extrapolation of β_0' to lower energies ($\varepsilon < 2.3$ eV). Subtracting from β_0' the extrapolated value of β_1', we find β_2'. Since β_0' increases very fast with decreasing ε for $\varepsilon < 2.3$ eV, the contribution of β_1' to β_0' will be comparable with the contribution of β_2' only in a very small energy interval. It is seen in Figure 32 that already at $\varepsilon = 2.0$ eV, β_1' becomes small compared with β_2'. For such a small energy interval, it is quite permissible to use the extrapolation of the β_1' (ε) curve.

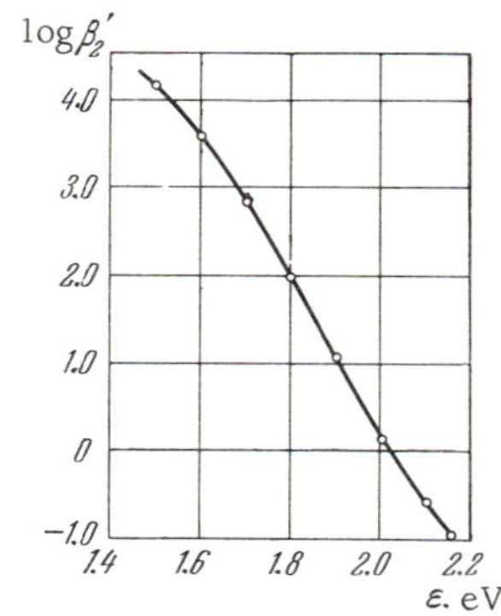

FIGURE 34. Dependence of log β_2'
on the quantum energy ε

The function β_2' obtained as the difference β_0' (ε) $- \beta_1'$ (ε) is shown in Figure 34 on semilogarithmic scale. It is seen that β_2' (ε) increases very fast with decreasing ε, varying by three orders of magnitude as ε changes by as little as $0.3\,\mathrm{eV}$. Near $\varepsilon = 1.5\,\mathrm{eV}$ the rate of growth of β_2' slows down and apparently, somewhere near $\varepsilon = 1.4$ eV, β_2' goes through a maximum. Unfortunately, the value of β_2 for $\varepsilon < 1.5\,\mathrm{eV}$ was not investigated in this sample. More detailed investigations of $\beta_2'(\varepsilon)$ in this energy interval were carried out later with other samples and it became clear that the maximum of $\beta_2'(\varepsilon)$ is at $\varepsilon \approx 1.32\,\mathrm{eV}$. These data, together with those of Figure 34, indicate that tunneling through the barrier at an energy level which we shall subsequently denote by $E_{C'}$ constitutes the second channel for capture of electrons by s-centers. To excite an electron to the level $E_{C'}$, an energy of $1.32\,\mathrm{eV}$ is required. Other data will also be presented later to confirm the existence of this trapping channel.

Thus the probability that an electron of energy $E_{C'}$ falls into a long-life trap is quite high. This is only possible if the level $E_{C'}$ is not a localized level but an extended energy level in which the electron can move all through the sample. Therefore, it may be assumed that $E_{C'}$ constitutes a narrow band of levels which appear in selenium exposed to mercury vapor. That the appearance of such a band of levels $E_{C'}$ is connected with the action of mercury vapor follows from the fact that in pure amorphous selenium films no such levels were observed /10/. Such films are not photosensitive in the range $\varepsilon = 1.3-1.4\,\mathrm{eV}$.

This impurity band, which we shall denote by C', is quite narrow. The half-width of the $E_{C'}$ level (determined at the half-height of the maximum) is $0.1-0.15\,\mathrm{eV}$. As we shall see later (Chapter 5, Sec. 3), the width of the C' band increases with the degree of mercury activation of the film, a characteristic feature of an impurity band.

It is also interesting to follow the variation with quantum energy of $K\beta\tau_0\gamma$. It follows from (3.34) that

$$K\beta\tau_0\gamma = \frac{1}{eu_pQ}\,\sigma_a(\lambda)\cdot S(\lambda) = f(\varepsilon). \qquad (3.39)$$

The factors $\sigma_a(\lambda)$, $S(\lambda)$, on the right in equation (3.39) are determined directly from experimental data; eu_pQ, as will be shown in Chapter 5, may also be calculated from data on conductivity variation during the cooling and heating of the sample. All these quantities were measured for a number of samples, and the curve in Figure 35 was constructed according to these data. The dependence of $K\beta\tau_0\gamma$ on ε is not a monotonic function and it has a distinct two-stepped character. For small energies $(\varepsilon < 1.8\,\mathrm{eV})$, the product $K\beta\tau_0\gamma$ increases very fast with photon energy. When ε goes from 1.0 to $1.4\,\mathrm{eV}$, this product increases almost by two orders of magnitude, following which the growth of log $(K\beta\tau_0\gamma)$ slows down and the curve $f(\varepsilon)$ clearly approaches saturation in the energy range $1.8-2.2\,\text{eV}$. However, when the photon energy is further increased, $\log(K\beta\tau_0\gamma)$ again markedly increases, approaching a new saturation level for $\varepsilon > 2.8\,\text{eV}$. The impression is that the curve $f(\varepsilon)$ consists of two distinct parts A and B, indicating the existence of two absorption bands in mercury-activated selenium films. The shape of the sections A and B is not unlike the spectral distribution of the absorption coefficient of amorphous selenium near the absorption edge. Figure 36 shows, on semilogarithmic scale, the curve of K vs. ε for amorphous selenium $/28/$. The position of the absorption edge determined from the width of the energy gap is marked on this curve by a vertical bar. Note

$$K\beta\tau_0\gamma = K\beta\tau_0 L\frac{\gamma}{L} = \Delta n\frac{\gamma}{L}$$

is proportional to the concentration of nonequilibrium carriers. If it is assumed that the B section of curve $\log K\beta\tau_0\gamma$ (see Figure 35) characterizes the spectral distribution of nonequilibrium carriers in the conduction band of activated amorphous selenium films exhibiting anomalous photoconductivity, then the A section characterizes the spectral distribution of carriers in the C' band of these films.

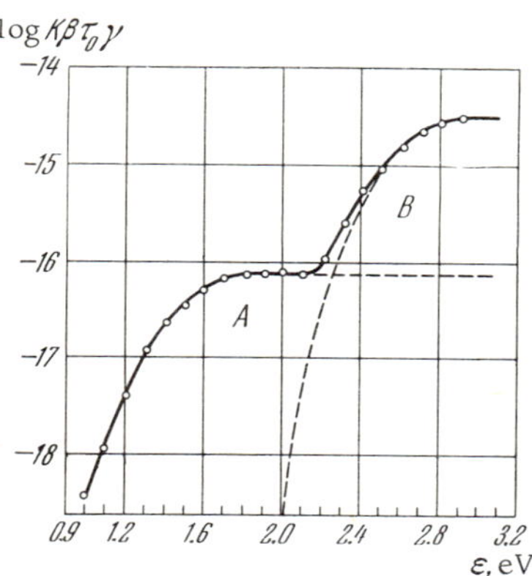

FIGURE 35. Dependence of log $K\beta\tau_0\gamma$ on photon energy ε

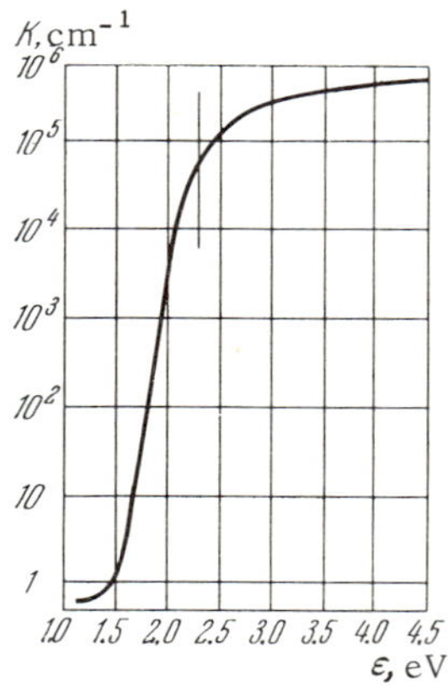

FIGURE 36. Dependence of the absorption coefficient K of amorphous selenium on photon energy ε

It was originally assumed /80/ that the C' band is situated in the energy gap of amorphous selenium $(E_g \simeq 2.4\,\text{eV})$ forming an impurity band. However, data considered in the next section support the assumption that electrons are not excited into the C' band from the valence band but from another band V' separated from the C' band by an energy $E \simeq 1.3\,\text{eV}$. Apparently, mercury-activated amorphous selenium films have two independent systems of levels, one of which (the usual system) is characterized by an energy gap $E_g \simeq 2.4\,\text{eV}$, while the other (C', V') has a smaller energy gap $E_g \simeq 1.3\,\text{eV}$. It is natural to assume that one of these systems is characteristic of the surface energy levels of a selenium film, while the other of the bulk energy levels.

Thus, at least two important conclusions follow from the data on the energy dependence of the quantum yield for electron capture by long-life traps.

1. Mercury-activated amorphous selenium films contain two systems of bands, the V, C (valence and conduction) bands, on the one hand, and the V', C' bands, on the other, with excitation energies of $2.4\,\text{eV}$ and $1.3\,\text{eV}$, respectively.

2. A minority carrier may penetrate into an s-center by two channels, namely: a) by tunneling through the barrier at the energy level corresponding to the bottom of the conduction band (tunnel effect); b) by tunneling through the barrier at the energy level $E_{C'}$. It is via this channel that electrons excited in selenium by photons of $\varepsilon < 2.0\,\text{eV}$ penetrate into s-centers.

8. THE EFFECT OF TEMPERATURE ON ANOMALOUS PHOTOCONDUCTIVITY

The influence of temperature on anomalous photoconductivity was investigated by several methods: 1) thermally stimulated currents were studied in samples of mercury-treated amorphous selenium; 2) the variation with temperature of the spectral distribution and the cross-section were studied; 3) prolonged observations were carried out of the dark conductivity at various temperatures; 4) conductivity curves under slow cooling and heating of the sample were studied.

The purpose of these investigations was, on the one hand, to determine the temperature dependence of the parameters characterizing anomalous photoconductivity and long-life traps and, on the other, to clarify the reasons for disappearance of anomalous photoconductivity at room temperature.

The results of the investigation of thermally stimulated currents during slow cooling and heating will be considered in Chapter 5. Here we shall present data concerning the influence of temperature on the spectral distribution of anomalous photoconductivity and the results of long-term observations of the steady-state level of dark conductivity at various temperatures.

a) The effect of temperature on the spectral distribution of anomalous photoconductivity

As temperature is varied, both the steady-state conductivity and the spectral distribution of $\sigma_a(\lambda)$ change /34/. These changes are illustrated in Figure 37. In this figure, the ordinate represents the steady-state conductivity established after switching off the illumination, and the abscissa gives the corresponding wavelength. The family of curves in this figure represents the spectral distribution of anomalous photoconductivity at various temperatures from 123° to 171°K. The measurements had to be confined to this narrow temperature interval because at higher temperatures irreversible processes take place in the sample and the spectral distribution becomes dependent on the order and duration of the measurements. In spite of the narrow range of temperature variation, it is clear that the spectral distribution changes with temperature. This change of the spectral distribution may be due to three factors:

1) changes with temperature in the concentration of metastable carriers generated by the filling of s-centers;

2) changes with temperature of the equilibrium concentration of carriers: the equilibrium carriers may contribute, though little, to the conductivity established in the dark, which is taken as the anomalous photoconductivity $\sigma_a(\lambda)$;

3) changes in the mobility of both the equilibrium and the metastable carriers.

All these changes can be easily separated taking into consideration the following points:

a) The component of the dark current due to equilibrium carriers should be independent of the spectral composition of the previous illumination and, therefore, its temperature dependence will be the same over the entire spectrum. Consequently, the difference $\sigma_a(\lambda_1) - \sigma_a(\lambda_2)$ will not contain the component which depends on the concentration of equilibrium carriers.

b) The variation of the mobility of metastable carriers with temperature will have the same effect on the anomalous component of photoconductivity irrespective of wavelength, so that $\dfrac{\sigma_a\,(\lambda_1)}{\sigma_a\,(\lambda_2)}$ must be temperature-independent.

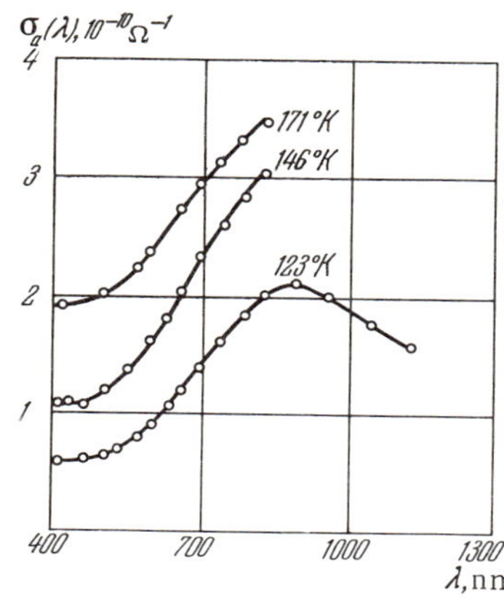

FIGURE 37.   Spectral distribution of the anomalous photoconductivity at various temperatures

The first step was accordingly to establish whether or not the ratio

$$\eta_i = \frac{[\sigma_a\,(\lambda) - \sigma_a\,(\lambda_0)]_{T=T_1}}{[\sigma_a\,(\lambda) - \sigma_a\,(\lambda_0)]_{T=T_i}} \tag{3.40}$$

is constant.

The invariability of this ratio with the wavelength λ would lead to the conclusion that the degree of filling of long-life traps does not change in the temperature interval in which this ratio is conserved and that consequently the temperature dependence of the difference $\sigma_a\,(\lambda) - \sigma_a\,(\lambda_0)$ is entirely due to the variation of the mobility of metastable carriers which give rise to the anomalous component of the photocurrent.

It turned out that in the relevant temperature interval, η_i was indeed independent of λ. Consequently, the temperature dependence of η_i is due to changes in the mobility of metastable carriers as a function of temperature, and therefore η_i is in fact nothing else than the ratio of these mobilities at temperature T_1 and T_i,

$$\eta_i = \frac{u\,(T_1)}{u\,(T_i)}. \tag{3.41}$$

The mobility of carriers constituting the anomalous component of the current (metastable carriers) does not vary markedly with

temperature.   Figure 38 shows the temperature dependence of the
mobility of metastable carriers.   The ordinate gives the ratio of
the carrier mobility at a given temperature to the mobility at
$T = 123°K$.   It is seen from Figure 38 that the mobility varies in
quite a complicated manner, which points to a complex scattering
mechanism of metastable carriers.   For the temperature interval
$123°-146°K$, the mobility grows with increasing temperature T as
$T^{5/2}$;   at higher temperatures it decreases with increasing
temperature.

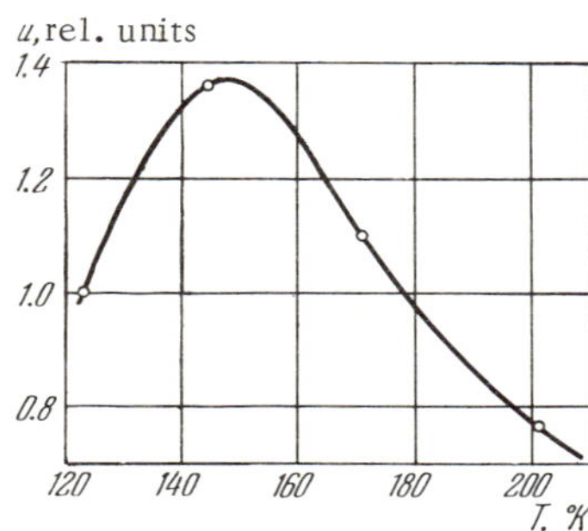

FIGURE 38.   Temperature dependence
of the mobility of metastable carriers
in mercury-activated amorphous selenium films

Further, from the shape of the curves shown in Figure 37, it may
be concluded that the ratio

$$Z = \frac{\sigma_a\,(700)}{\sigma_a\,(400)}$$

undergoes large changes with temperature.   It has been established
above (and will be confirmed subsequently) that the concentration of
filled s-centers does not change in the given temperature interval
(η_i does not depend on λ), and therefore the variation of Z with
temperature indicates that the conductivity established in the dark
$\sigma_a\,(\lambda)$ is due not only to the metastable but also to the equilibrium
carriers, i. e.,

$$\sigma_a\,(\lambda) = \sigma_0 + \sigma_m\,(\lambda).$$

$\sigma_m\,(\lambda)$ and σ_0 have different temperature dependences.
It is evident that the relative contribution of σ_0 to the magnitude
$\sigma_a\,(\lambda)$ will be larger the smaller $\sigma_m\,(\lambda)$.   The smallest value of
$\sigma_m\,(\lambda)$ appears in the short-wave region of the spectrum ($\lambda = 400\,\text{nm}$).
With increasing temperature, $\sigma_a\,(400)$ grows fastest, whence it fol-
lows that σ_0 increases with temperature faster than $\sigma_m\,(\lambda)$.   Knowing

how $\sigma_m(\lambda)$ varies with temperature, it is possible to determine the temperature dependence of σ_0. σ_0 is found to vary exponentially with temperature, with an activation energy $\Delta E = 0.16\,\text{eV}$, i.e.

$$\sigma_0 \sim e^{-\frac{0.16}{kT}}. \tag{3.42}$$

This is, however, also the activation energy of hole mobility in amorphous selenium /12, 26/. It is also known /35/ that in amorphous selenium the concentration of equilibrium carriers does not depend on temperature and the temperature dependence of the dark conductivity σ_d is determined only by the variation of mobility. This suggests that the dark conductivity of the nonanomalous component in selenium is hole conductivity in the valence band.

The different temperature dependences of the mobility of the anomalous and the nonanomalous current components show that the corresponding mechanisms are different. In amorphous selenium, the electron mobility varies exponentially like the hole mobility, but with an activation energy of $0.25\,\text{eV}$ (exceeding the activation energy for holes, $\Delta E = 0.16\,\text{eV}$). The fact that the temperature coefficient of the mobility of metastable carriers is small (considerably smaller than for electrons and holes in selenium) indicates that anomalous photoconductivity in amorphous selenium is neither conductivity in the conduction band nor conductivity in the valence band. Anomalous photoconductivity is thus probably surface conductivity, whose magnitude should be quite substantial in mercury-activated amorphous selenium films.

b) The effect of temperature on the parameters of s-centers

According to /27/, the s-centers should retain the properties of long-life traps at room temperature. This conclusion is based on the assumption that the parameters of s-centers do not change with temperature. To prove this assumption, the effect of temperature on the spectral distribution of the cross-section S was investigated in /36/. Figure 39 shows the results of these measurements (sample No. 7VK). In this figure the abscissa represents the photon energy of the incident light, and the ordinate is the logarithm of the cross-section. Curve 1 in Figure 39 plots $\ln S(\varepsilon)$ at $T_1 = 123°\text{K}$, curves 2 and 3 correspond to $\ln S(\varepsilon)$ at $T_2 = 171°\text{K}$ and $T_3 = 186°\text{K}$. As the temperature varies, $S(\varepsilon)$ undergoes some changes. From the shape of these curves in the interval $\varepsilon = 2.3-3.0\,\text{eV}$, the values of the parameters r_0, U_0, E_b were calculated, and are shown in Table 4.

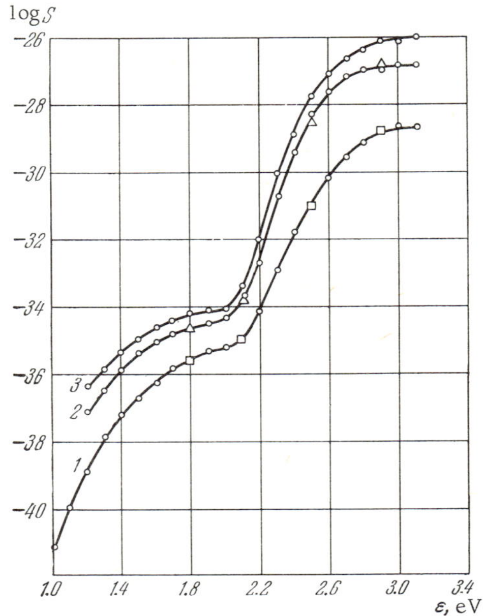

FIGURE 39. Effect of temperature on S (ε)

It follows from Table 4 that the parameters r_0, U_0 do not undergo substantial changes with temperature. The scatter of the data lies within the limits of experimental error. The variation of the parameter E_b, on the other hand, exceeds the limits of experimental error. However, the absolute value of the change in E_b is insignificant and cannot affect considerably the escape probability of a carrier from an s-center. Therefore, the conclusion of /27/ that the s-center remains a long-life trap at room temperature is still valid.

TABLE 4

Parameters Temperature	U_0, eV	r_0, A	E_b, eV
123 °K	3.5±0.4	51±7	2.96±0.01
171 °K	3.2±0.2	50±7	2.84±0.01
186 °K	3.05±0.15	53±7	2.82±0.02

c) Long-term observation of the steady-state dark conductivity at various temperatures

In order to clarify how the degree of filling of long-life traps varies with temperature, prolonged observations of the dark conductivity were carried out at various temperatures /37/. The sample exhibiting anomalous photoconductivity was illuminated at $\lambda_1 = 740\,nm$. This wavelength was chosen in order to produce a substantial filling of traps under illumination. The illumination was then switched off and the sample left in the dark until its conductivity reached the steady-state level of anomalous photoconductivity $\sigma_a (740)$. The constancy of the conductivity value was verified by observation during $\simeq 20\,hr$. The temperature was then raised. As the temperature of the sample was increasing (during $1-1.5\,hr$), its resistance changed. After the establishment of the new temperature, conductivity measurements were repeated for $10-20\,hr$ so as to clarify whether the sample conductivity remained time independent at the new temperature also. The constancy of conductivity would indicate that the degree of filling of long-life traps remained unchanged at the new temperature. In this case, it could be assumed that the concentration of filled long-life traps also remained constant during the process of temperature variation, and that the observed change in conductivity was due to a change in the mobility of metastable carriers and the nonanomalous component of conductivity.

The results of the measurements are shown in Figure 40. The abscissa gives the length of observation t in hours and the ordinate gives the conductivity σ_d. The zero point of the time axis corresponds to the beginning of the fixed temperature phase.

Figure 40 shows that the conductivity remains constant in time in the entire temperature interval from 123° to 170°K. Only at a temperature of 177−180°K does a slow change in conductivity begin to be observed, at a rate of the order of $10^{-13}\,\Omega^{-1}/hr$, which constitutes approximately 0.1% σ_d per hour. It is significant that the conductivity increases, and not decreases, in time. As the temperature is raised, the rate of conductivity change grows and at 202°K it reaches $6 \cdot 10^{-12}\,\Omega^{-1}/hr$. Note that a change in conductivity by as much as $6 \cdot 10^{-11}\,\Omega^{-1}$ (the change of σ_d over 10 hr at 202°K), which constitutes $\sim 16\%$ of the initial value of σ_d, does not affect the magnitude of $d\sigma/dt$ noticeably. This indicates that the concentration of filled s-centers q is still small compared with their total concentration Q.

It could be assumed that the increase of the dark conductivity observed in activated selenium films at certain temperatures is a phenomenon related to thermally stimulated currents (TSC).

However, in activated selenium films, this growth of conductivity has a number of specific features. No momentary TSC peaks are observed. The increase of conductivity is a slow process, taking perhaps hours and days depending on the temperature, until a definite constant value is reached which is preserved for an indefinite length of time. The process of conductivity increase is characterized by a definite activation energy. Figure 41 plots the logarithm of the rate of conductivity increase $\ln \dfrac{d\sigma_T}{dt}$. It is seen in the figure that this dependence is linear. From the slope of the straight line, the activation energy of thermally stimulated conductivity growth in activated amorphous selenium films was determined. It was found to be $0.6 \pm 0.1\,\mathrm{eV}$.

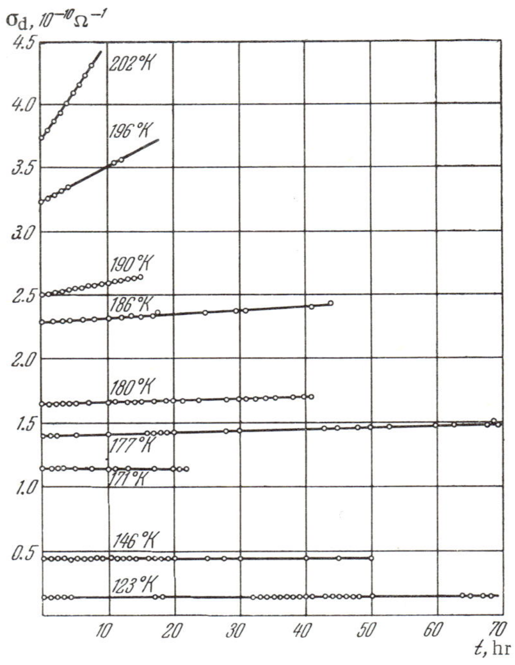

FIGURE 40. The effect of temperature on the kinetics of dark conductivity in anomalously photoconductive selenium (sample No.7VK)

The increase of the dark conductivity at constant temperature may be due to either of two processes.

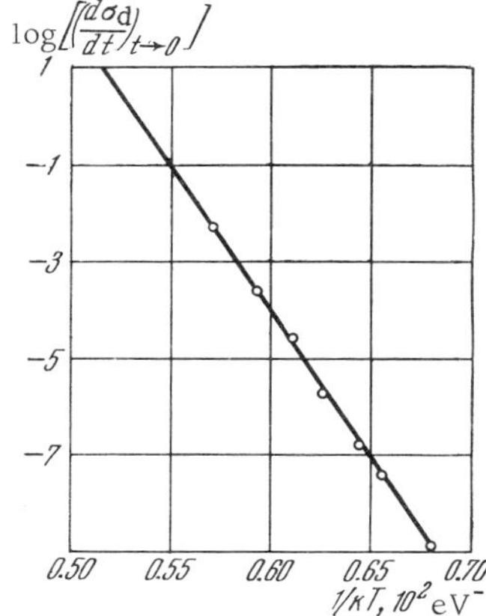

FIGURE 41. Temperature dependence of the logarithm of the rate of filling of s-centers by carriers. The rate of filling is expressed in units of $10^{10}\,\Omega^{-1}$/hr.

1. Liberation of majority carriers captured by traps during illumination at a lower temperature.

2. Continuous capture of minority carrier by traps, as a result of which the equilibrium is disturbed and the concentration of majority carriers increases.

The first process occurs usually with thermally stimulated currents. It takes a comparatively short time. The increased conductivity persists for a time interval determined by the rate of recombination of the liberated carriers. It is evidently not this process that takes place in activated selenium films. We already know that the removal of a carrier from an s-center requires a much larger expenditure of energy, around 3 eV, and a process with an activation energy of 0.6 eV thus cannot be responsible for freeing carriers from s-centers. According to data presented in Sec. 6, a carrier captured by an s-center is also stored for an indefinite length of time at temperatures higher than 200°K. In the dark at temperatures below 200°K, a carrier cannot escape from an s-center.

The thermally stimulated increase of conductivity in activated selenium films proceeds in such a way that the generated majority carriers are shielded from recombination, for these films preserve the increased conductivity for an indefinite length of time. Therefore, the thermally stimulated increase of anomalous photoconductivity in mercury-activated selenium films does not liberate thermal carriers from s-centers. If this is so, then this must be a process of carrier capture by s-centers. Carriers captured by s-centers are stored for an indefinitely long time. Since the conductivity increases as a result, we conclude that the s-centers in selenium capture minority carriers. The free majority (metastable) carriers arising as a result of the capture of minority carriers by s-centers have no partners to recombine with, so that the change in conductivity has a residual character and will persist for an indefinite length of time.

The activation energy of $0.6 \pm 0.1\,$eV agrees well, within the limits of experimental error, with the value of $E_{C'}/2 = 0.66\,$eV. If this agreement is not accidental, it can be assumed on the basis of these data that 1) the s-centers are filled in the dark by equilibrium carriers through the C'' band, and 2) in the dark, at sufficiently high temperatures, all the s-centers are eventually filled.

Since the activation energy of $d\sigma_d\,dt$ is known, it is possible, taking the value of $d\sigma_d/dt$ at some given temperature, to calculate its value at any temperature and, in particular, to determine whether $d\sigma_d\,dt$ between $120-160°$K is indeed so small that it can be neglected.

The results of such a calculation are presented in Table 5. It is seen that at low temperatures $d\sigma_d/dt$ is indeed very small.

TABLE 5. Rate of change of the anomalous component of conductivity at various temperatures as a result of the filling of s-centers

Temperature, °K	120	130	140	150	160	180
$\dfrac{d\sigma_d}{dt}$, Ω^{-1}/hr	$4.3 \cdot 10^{-22}$	$3.8 \cdot 10^{-20}$	$2.2 \cdot 10^{-18}$	$3.6 \cdot 10^{-17}$	$7.2 \cdot 10^{-16}$	10^{-13}

Thus at $120°$K, $d\sigma_d\,dt = 4.3 \cdot 10^{-22}\,\Omega^{-1}$/hr.

The minimum change of anomalous conductivity detectable in these experiments $(0.1\%\ \sigma_d)$ was $10^{-13}\,\Omega^{-1}$. Consequently, it would take more than $2 \cdot 10^8$ hr, or $3 \cdot 10^4$ years, to detect, at this temperature, a change in conductivity. Even at $160°$K, about 150 hr (or more than 6 days) would be required for the conductivity σ_d to change by 0.1%.

The data of Table 5 show that the assumption concerning the absence of thermal excitations into s-centers in activated selenium at temperatures $120-160°$K is indeed true.

Thus, disappearance of anomalous photoconductivity in mercury-treated, amorphous selenium layers is due to the increase of the free electron concentration in the C' band. The nonanomalous component of conductivity is thereby increased and the rate of leakage of electrons from the band into the long-life traps becomes appreciable.

9. THE PHYSICAL NATURE OF LONG-LIFE TRAPS

As mentioned before, no anomalous photoconductivity is observed in amorphous selenium layers not treated with mercury vapor. Consequently, these layers do not contain long-life traps. The latter appear after the mercury vapor treatment. It is known /7/ that the long-life traps are located at the bottom of "valleys" of the surface microrelief of the selenium layers, either on the surface itself or in a subsurface layer. The development of long-life traps following mercury vapor treatment indicates that the material of long-life traps contains mercury. It is possible that the whole trap consists of mercury, i.e., a mercury drop, but it is more likely that the trap consists of mercury selenide or of some chemical compound in which mercury, and possibly also oxygen, take part. Such drops (of mercury or mercury selenide) may form on the surface, in the valleys of the microrelief, or in free cavities in the subsurface layer.

The existence in nonmetallic crystals of colloidal metal particles of bulk or impurity origin has been known for a long time. However, not much is known about subcolloidal particles, i.e., embryonic nuclei which grow and develop into colloidal metal particles. Only recently have some models and theoretical ideas on the fluctuational origin of these centers been formulated. The first fluctuational model, applicable to alkali halide crystals, was suggested in /38/. There it was shown that the self-localization of electrons at discrete levels in a potential well of fluctuational origin leads, under certain conditions, to such an energy gain that these quasi-metallic centers (QMC, to use the terminology of A. E. Glauberman) become thermodynamically favorable, i.e. correspond to a minimum of the thermodynamic potential of the system. It is also suggested in /38/ that long-life traps may be identified with such QMC. In /39, 40/ the theory of optical and photoelectric properties of QMC with dimensions of the order $10-100\,\text{Å}$ is developed. The special form of the fine structure observed under certain conditions in the absorption spectra of silver halide /41/ was successfully explained. We did not observe such a fine structure in the absorption spectrum of activated amorphous selenium layers. This may be due to the comparatively large size of s-center particles in these films. According to /39/, the distance between the fine structure components varies in inverse proportion to the cube of the particle size, and for particle size of $\sim 70\,\text{Å}$ it becomes so small that the fine structure disappears. It is, however, possible that, because of their low concentration, the absorption due to the subcolloidal particles is not noticeable against the absorption background of the selenium film, which is quite substantial in the range $400-700\,\text{nm}$.

It was shown in /42/ that in disordered systems (including amorphous selenium films) stable macroscopic formations, so-called "fluctuons," may appear as a result of the nonlinear dependence of the thermodynamic potential on the corresponding internal parameter for a sufficiently large ratio of the binding energy to kT. In s-centers of amorphous selenium this condition is fulfilled, $E_b/kT > 100$. The fluctuons are localized near the second (nonequilibrium at this temperature) phase. In some cases (for small heats of transition) localized electron states may exist in a comparatively wide temperature interval. In amorphous selenium films, there are always small regions of crystalline phase, and it is therefore not excluded that the s-centers in activated selenium films are precisely of this "fluctuon" origin. It should be noted that no s-centers are observed in nonactivated selenium films, while in activated films the s-centers appear only in cavities on the surface of the selenium layer. It is therefore also possible that the s-centers in selenium are of a different origin, being associated with the surface states of these films and the fairly high electron affinity of the surface states.

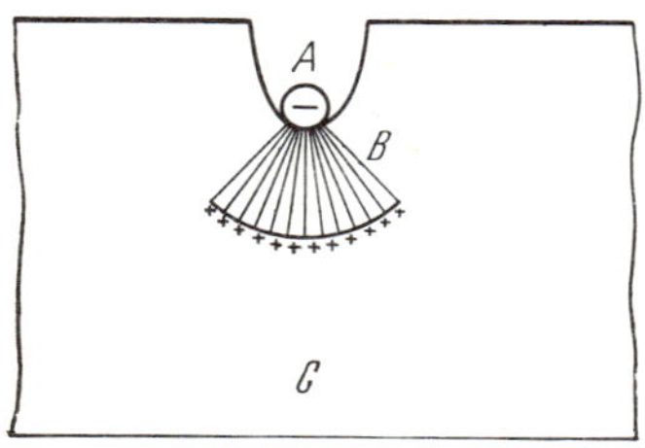

FIGURE 42. Schematic diagram of the physical model of a long-life trap

Let us assume that on the surface of the above-mentioned cavity there are surface energy-levels of the Tamm-level type, associated with the boundary of the solid or with the presence on the surface of adsorbed atoms which have a definite electron affinity. Such energy levels will capture electrons which either exist in selenium or are produced in it by absorption of light. As a result, the selenium surface becomes negatively charged. The magnitude of the electron charge q formed on the selenium surface is determined by the electron affinity U_a of the surface centers. The charging buildup on the surface will continue until the work necessary to bring one electron to the surface reaches the value of U_a

We shall further assume that the subsurface selenium layer is polarized, i.e., the long selenium chains are oriented perpendicular to the surface, that these chains are fairly long (over 70 Å), and that the ends of the chains constitute hole traps, so that the entire positive charge corresponding to the negative charge bound in the surface centers is concentrated at the chain ends. In Figure 42, these chains are identified by the letter B; the drop whose surface is negatively charged is identified by the letter A, and the selenium layer by the letter C. As a result, a sort of condenser is formed in the selenium layer with charged surfaces of nearly spherical shape. In this condenser, the field decreases as $1/r^2$, i. e., for an electron in selenium this field will constitute a Coulomb barrier of height $U_0 = q/r_0$, where q is the charge on the minor surface of the condenser.

It is further assumed that the material of the trap (mercury or mercury selenide) is a good conductor and that the long-life trap itself constitutes a potential well for the electron. The binding energy of an electron in a trap is evidently the work function of the electron in the trap material. This quantity must be approximately constant, varying somewhat as a function of the radius of curvature r_0 of the drop.

U_0 depends on the parameters of the spherical condenser and on the affinity U_a, namely

$$U_a = q\left(\frac{1}{r_0} - \frac{1}{R}\right) = \frac{q}{r_0}\left(1 - \frac{r_0}{R}\right) = U_0 \frac{1}{1 + r_0/l},$$

or

$$U_0 = U_a (1 + r_0/l), \qquad\qquad (3.43)$$

where l is the length of the selenium chains associated with the long-life trap, and $R = r_0 + l$.

It is seen from relation (3.43) that the height of the potential barrier surrounding a trap is mainly determined by the electron affinity of the surface centers, although it is somewhat greater than the affinity. A change of the chain length will slightly affect the value of U_0. Note that when the condition $l > r_0$ is fulfilled, it is not necessary to assume that the lengths of all the chains are the same.

This model of a long-life trap explains the small scatter in the values of principal trap parameters. The model is in agreement with the view that the long-life traps are situated in "valleys" of the surface microrelief of selenium films, and it assigns a clear role to mercury in the formation of long-life traps.

10. SLOW RELAXATION IN CdS AND SOME OTHER MATERIALS

Very long relaxation times, so-called "frozen conductivity," have lately been reported /43—53/, which appear in semiconductors at low temperature as a result of exposure to light. The phenomena were observed in CdS single crystals /43/, polycrystalline CdS films /44—46/, in the CdS—SiO$_x$ film system /49/, in polycrystalline layers of AgI and ZnO /50/, in glassy Tl$_2$Se·As$_2$Te films /51/ and in thin films of some organic dyes /52/.

The term "frozen conductivity" implies that the conductivity excited in the sample by exposure to light is preserved in the dark. This phenomenon is observed at low temperatures (whence the term "frozen conductivity"), though in some semiconductors /52/ it is also observed at room temperature. Since the same features are also inherent in anomalous photoconductivity, there arises the question as to the relation between the two phenomena.

On the whole, the investigation of frozen conductivity is not complete enough, the mechanism of the phenomenon has not been fully clarified, and different authors have advanced different hypo- theses to explain this phenomenon. Thus in /53—54/, the large relaxation time (frozen conductivity) is attributed to the intermittent movement of carriers as a result of their capture by shallow traps, to multiple recapture of carriers /52/, or to a barrier for recom- bination /49, 51/.

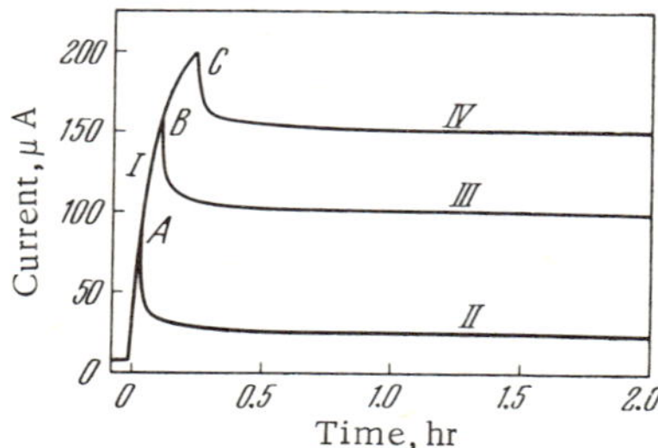

FIGURE 43. Photocurrent relaxation curves in CdS layers

Anomalous photoconductivity in mercury-treated selenium samples is also due to the capture of carriers by long-life traps (s-centers) which constitute a potential well surrounded by a wide Coulomb potential barrier.

However, there are a number of very substantial differences be- tween the frozen conductivity in these materials and anomalous photoconductivity. To illustrate these differences, we shall consider the data of /45/ on slow relaxations in CdS films.

Figure 43 shows conductivity relaxation curves under illumination and in the dark in polycrystalline CdS layers at liquid helium temperature /45/. The relaxation curves in Figure 43 plot the time dependence of the current through the sample.

Curve I shows the relaxation of the current through a sample illuminated with a light flux of a certain intensity. The illumination could be switched off at a fixed instant (points A, B, C) and the variation of conductivity was observed thereafter (curves II, III, IV). After switching off the illumination, a relatively fast drop in the magnitude of the current was first observed followed by a slow relaxation. The relaxation time was found to depend on the intensity of the light pulse incident on the sample: the higher the intensity, the longer the relaxation time. The time required for the light-induced conductivity to decrease to its half-value was determined from a 10-hour observation of conductivity variation. At 4.2°K, this time was 100 hr, at 77°K 24 hr, and at 300°K 2 hr. The initial dark value of conductivity could be rapidly restored by heating the sample.

Similar evaporated CdS films were also investigated in our laboratory /47/. The experiments were conducted at liquid nitrogen temperature. The level of frozen conductivity was found to depend on light intensity and there was no spectral memory in CdS samples at liquid nitrogen temperature. When illuminated with red and blue light, the sample relaxed, though slowly, to the same initial level of dark conductivity, i.e., in these samples we are not dealing with long-life traps but with traps surrounded by a comparatively low barrier, which are the cause of the slow relaxation.

The temperature dependence of σ supports this suggestion. Thus, when the temperature was varied from 77 to 300°K, the conductivity relaxation time changed by a factor of 12, which corresponds to a barrier $\Delta E = 0.02$ eV. It is clear that no spectral memory can occur at such small barriers.

It is not excluded that long-life traps (centers with a wide potential well surrounded by a high Coulomb barrier) may be responsible for slow relaxation in some materials. To clarify this, prolonged investigations are necessary to determine the carrier lifetime in these materials, the presence of spectral memory, the energy dependence of the cross-section, none of which is sufficiently covered in the literature.

Let us summarize the results of Chapter 3.

1. It has been established experimentally that at temperatures at which anomalous photoconductivity is observed in amorphous selenium films, the rate of thermal excitation of carriers into s-centers is negligibly small and only at $T \simeq 180°K$ does this mechanism of carrier excitation begin to become noticeable.

2. The s-centers are formations containing many thousands of atoms (the subcolloidal model).

3. The potential function of an s-center constitutes a wide potential well $(a_0 \simeq 70\,\text{Å})$ surrounded by a potential barrier, tapering toward its top. The barrier height is $U_0 \simeq 3\,\text{eV}$, and the electron binding energy $E_b \simeq 2.9\,\text{eV}$.

4. The s-centers are situated in valleys of the surface relief of the film.

5. In all the amorphous selenium films investigated, the s-centers have practically the same potential function with very close values of the parameters $(a_0,\ U_0,\ E_b)$.

6. An s-center possessing the above-mentioned properties is a long-life trap not only at low temperatures but also at room temperature.

7. The subcolloidal model of s-centers accounts for all the known properties of anomalously photoconductive selenium.

Chapter 4

NORMAL PHOTOCONDUCTIVITY IN SEMICONDUCTORS CONTAINING S-CENTERS

1. PHOTOCONDUCTIVITY OF MERCURY-ACTIVATED AMORPHOUS SELENIUM FILMS NEAR ROOM TEMPERATURE

Anomalous photoconductivity in mercury-activated amorphous selenium films is observed only at temperatures below 200°K. At room temperature or near it, photoconductivity in these films is substantially different. They have no spectral memory. After the illumination is switched off, the conductivity returns, in a relatively short time, to the same σ_d, irrespective of the spectral composition of illumination. In these conditions, the dark conductivity thus becomes the characteristic electrical property of the sample. The conductivity produced by illumination is a single-valued function of light intensity, which makes it possible to use the notion of photoresponse in this case.

Therefore, following the definition given in Sec. 1, Chapter 1, we must characterize such conductivity as normal.

Photoconductivity at room temperature in mercury-activated amorphous selenium films was investigated in a number of papers /6, 55—58/. The kinetics of this photoconductivity is very complicated and largely depends on the intensity and spectral composition of the radiation.

For the sake of illustration, Figure 44 shows typical relaxation curves for the films as illumination is switched on and off /6/. In some cases (Figure 44a), the sample conductivity increases upon switching on the illumination (positive photoconductivity), but only for a relatively short time: further illumination makes it go through a maximum and it starts to fall. A peculiar s e l f - q u e n c h i n g o f p h o t o c o n d u c t i v i t y t a k e s p l a c e. As a result, the steady-state conductivity σ_s is less than the maximum value σ_{max} observable during the relaxation process, and consequently the photoresponse $\Delta\sigma_{ph}$ is also less than the maximum change in conductivity during illumination $\Delta\sigma_{max}$. When the illumination is switched off, the conductivity decreases and returns to its initial dark value (Figure 44a). Sometimes two distinct sections are apparent on the relaxation curve after switching off the illumination: a section of fast decrease and a section of slower relaxation. In some cases the fast decrease is quite large and the conductivity quickly falls below its dark value (negative conductivity), returning to its dark value σ_d from the direction of negative $\Delta\sigma$ values (Figure 44b).

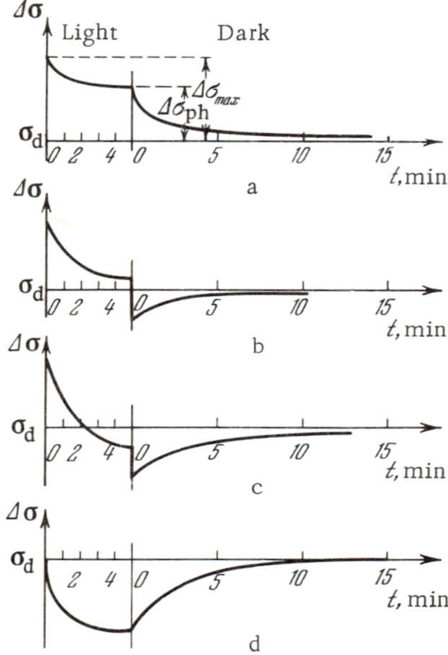

FIGURE 44. Various types of photoconductivity relaxation curves observed in mercury-activated amorphous selenium films at room temperature

In some cases the positive photoconductivity is relatively large (Figure 44c), in others it is altogether absent (Figure 44d), but the fall of conductivity which follows its increase is so large that the steady-state photoconductivity is not only less than the maximum value but even less than the initial dark conductivity. The photoresponse in this case is negative (negative photoconductivity).

These features create an impression that two kinds (two components) of photoconductivity are simultaneously present in the relaxation curves of activated selenium films, a positive and a negative one, with different magnitudes of the photoresponse and different relaxation times. We shall subsequently denote the photoresponse of the positive component by $\Delta\sigma^+$, and the photoresponse of the negative component by $\Delta\sigma^-$. The resultant photoresponse is $\Delta\sigma_{ph} = \Delta\sigma^+ - \Delta\sigma^-$. The relaxation times of the two photoconductivity components will be denoted by τ^+ and τ^-, respectively.

In different samples, for given illumination characteristics, any of the different cases of Figure 44 may be observed. However, all the photoconductivity relaxation patterns shown in this figure may occur in the same sample for different illumination characteristics. Curves a and b are observed when activated selenium films are illuminated with long-wave light $(\lambda > 580\,\mathrm{nm})$, curves c and d are observed when the films are illuminated with short-wave light $(\lambda < 500\,\mathrm{nm})$. Curves a and b occur at low light intensities, curves c and d at high intensities. Under white-light illumination, all the photoconductivity relaxation patterns shown in Figure 44 are encountered.

The photoresponse changes with intensity. Figure 45 shows conductivity relaxation curves for various light intensities at $\lambda = 730\,\mathrm{nm}$. If the illumination intensity for curve 1 is taken as unity, for curves 2, 3, 4, and 5 it is expressed by 0.805, 0.431, 0.194, and 0.055, respectively. Figure 46 shows conductivity relaxation curves for various intensities of short-wave illumination $(\lambda = 415\,\mathrm{nm})$. For curves in Figure 46, the intensity ratio $L_1 : L_2 : L_3$ is 1:0.39:0.025. In either case photoresponses $\Delta\sigma_{ph}$, $\Delta\sigma^+$, and $\Delta\sigma^-$ and the steady-state conductivities vary with intensity.

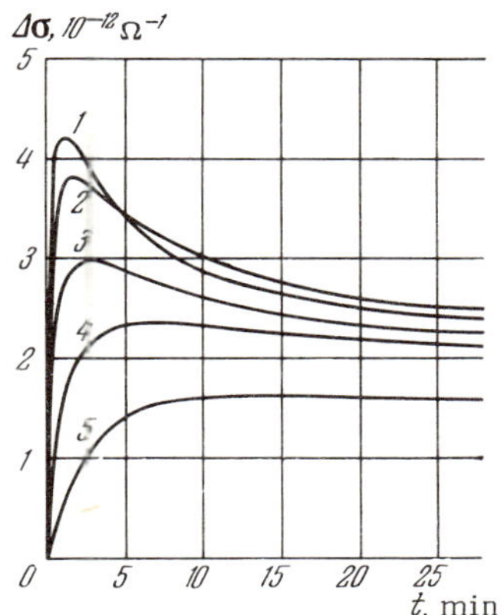

FIGURE 45. Photoconductivity relaxation curves in activated amorphous selenium films (sample No.10VK) for various light intensities at $\lambda = 730\,\mathrm{nm}$

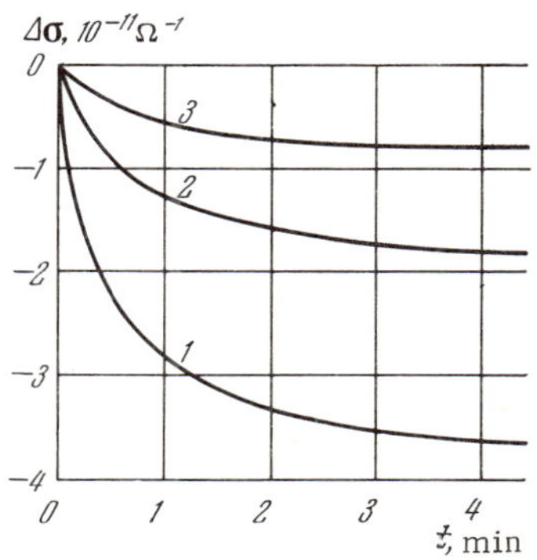

FIGURE 46. Photoconductivity relaxation curves (sample No.10VK) for various light intensities ($\lambda = 415\,\mathrm{nm}$)

The mechanism of the development of photoconductivity of such a complex character in activated amorphous selenium films has so far remained without explanation. In some investigations /6/, it was suggested that the mechanism of negative photoconductivity advanced by Stockmann /59/ could be applied to our case. According to

Stockmann, negative photoconductivity appears when photons create in the semiconductors mainly minority carriers, which recombine with thermally generated majority carriers.

The corresponding transitions are illustrated in Figure 47. Such a photoconductivity can be realized only under certain restrictive assumptions, namely:

1) the sample contains at least two types of localized levels I and II (Figure 47), with different excitation energies E_1 and E_2;

2) the Fermi level E_F falls between these levels;

3) there is a definite relaxation between the rate of thermal excitation of electrons from levels II into the conduction band and the rate of electron-hole recombination;

4) free holes must not recombine directly with electrons bound in levels II;

5) the capture cross-section of majority carriers for levels II must be much less than the capture cross-section for levels I;

6) the concentration of centers with energy levels E_1 and their cross-section for the capture of majority carriers are not very small.

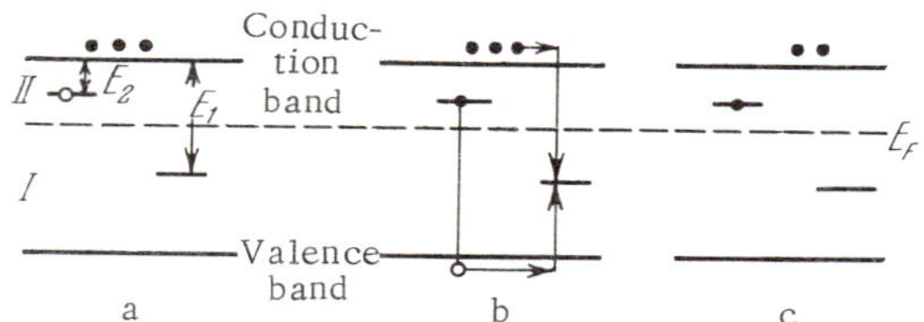

FIGURE 47.   The origin of negative photoconductivity:

a) carrier distribution before illumination (dots represent electrons, circles holes);   b) creation of holes under illumination;   c) steady state.

Since it is necessary that the majority of these conditions be fulfilled simultaneously, such photoconductivity will appear extremely rarely.

However, it is not shown in /6/ that the above requirements are fulfilled in mercury-activated amorphous selenium films, and therefore the question of the applicability of Stockmann's mechanism of negative photoconductivity remains open. In our opinion, since such films exhibit anomalous photoconductivity at low temperatures and contain s-centers, all the special features of photoconductivity in these films observed at room temperature, i. e., negative photoconductivity in blue light, self-quenching of photoconductivity in red light, must be associated with the presence of s-centers in activated films. Therefore, the following questions naturally arise.

1. Under what conditions can the photoconductivity of a semiconductor containing s-centers assume normal character?

2. Is simple heating of the sample sufficient for this purpose?

3. What is the character of this normal photoconductivity? Can it be the photoconductivity observed in activated amorphous selenium films at room temperature?

2. TYPES OF NORMAL PHOTOCONDUCTIVITY IN A SEMICONDUCTOR WITH S-CENTERS

In Chapter 2, while considering the problem of photoconductivity in a semiconductor with s-centers, we clarified the conditions under which anomalous photoconductivity may arise. It is clear that when these conditions are not met, the character of photoconductivity changes: the photoconductivity becomes normal.

The photoconductivity in a semiconductor will have an anomalous character when the concentration of nonequilibrium carriers is moderate and the concentration of equilibrium carriers is negligibly small. Such photoconductivity indeed exists at temperatures of 120—170°K in mercury-activated amorphous selenium films. However, when the temperature is raised to 200—300°K, the conductivity in these films acquires a normal (in the sense of its dependence on light intensity), but highly unique character.

What happened with the increase in temperature?

It follows from the data in Sec. 9, Chapter 3, that the s-center parameters, i. e., the width of the potential well, the height of the potential barrier, the carrier binding energy in an s-center, do not undergo substantial changes with the increase in temperature. In addition, in Sec. 6, Chapter 3, it was shown that despite the large increase with temperature of the probability of spontaneous liberation of a carrier from an s-center, the absolute value of this probability remains so small ($\theta \sim 10^{-18}$ sec), and the corresponding lifetime so large ($\tau_q \sim 10^{16}$ years), that the s-center can accordingly be considered a long-life trap with zero recombination cross-section for trapped carriers with free carriers.

The only change that should be taken into account when the temperature is raised is the increase in the concentration of electrons in the C' band and, as a result, the increase in the rate of filling of the free s-centers with carriers.

We shall now consider the changes in photoconductivity produced by this factor.

The kinetic equations which determine the relaxation of photo-conductivity can be written in the same form as before (Sec. 2, Chapter 2),

$$\frac{dn}{dt} = K\beta L - \frac{n}{\tau_0} + SqL - \gamma n(Q - q) + g,$$

$$\frac{dq}{dt} = \gamma n(Q - q) - SqL,$$

$$p = n + q, \quad g - \frac{n_0}{\tau_0} = 0, \tag{4.1}$$

with the only difference that the rate of equilibrium carrier generation g is no longer negligible and therefore the concentration of equilibrium carriers is not negligible in comparison with the concentration of nonequilibrium carriers. The total concentration of free electrons n is thus expressible as the sum $n_0 + \Delta n$.

When $\tau_0 \ll \tau(\lambda)$, where $\tau(\lambda)$ is the relaxation time for the filling of s-centers, the concentration of nonequilibrium carriers is as before

$$\Delta n = K\beta L \tau_0.$$

After switching off the illumination, a quasi-equilibrium state is first established, which can be characterized by zero nonequilibrium carrier concentration; a much slower process then follows, in which the s-centers are gradually filled and the concentration of metastable majority carriers $p_m = q$ is established. In steady state, all s-centers are filled and the concentration of metastable majority carriers is equal to the concentration of s-centers Q. The contribution of these carriers to the dark conductivity is $eu_m Q$.

Under illumination, nonequilibrium carriers are generated in the semiconductor, and simultaneously emptying of s-centers takes place. If, as before, $\tau_0 \ll \tau(\lambda)$, a quasi-steady state is first established, in which the concentration of nonequilibrium carriers remains practically unchanged,

$$K\beta L - \frac{n}{\tau_0} + g = 0 \tag{4.2}$$

and

$$\Delta n = K\beta L \tau_0,$$

whereas the concentration of metastable carriers has not yet undergone substantial changes. The relaxation of metastable carriers will take much longer and will be described by the equation

$$\frac{dq}{dt} = \gamma(n_0 + \Delta n)(Q - q) - SqL =$$
$$= K\beta L\tau_0\gamma(Q - q) - SqL + \alpha(Q - q), \qquad (4.3)$$

where

$$\alpha = \gamma n_0. \qquad (4.4)$$

It follows from (4.3) that

$$\Delta q = Q - q = (Q - q_{ss})\left(1 - e^{-\frac{t}{\tau(\lambda)}}\right) = \Delta q_{ss}\left(1 - e^{-\frac{t}{\tau(\lambda)}}\right), \qquad (4.5)$$

where q_{ss} is the steady-state concentration of filled s-centers under illumination, $\tau(\lambda)$ is the relaxation time for the establishment of photoconductivity when the semiconductor is illuminated with light of wavelength λ,

$$\tau(\lambda) = \frac{1}{\gamma K\beta L\tau_0 + SL + \alpha}. \qquad (4.6)$$

The steady-state values of q, p, n are determined by the relations

$$q_{ss} = \frac{(\alpha + \gamma K\beta L\tau_0)Q}{\alpha + \gamma K\beta L\tau_0 + SL},$$
$$n_{ss} = n_0 + K\beta L\tau_0, \qquad (4.7)$$
$$p_{ss} = n_0 + K\beta L\tau_0 + \frac{(\alpha + \gamma K\beta L\tau_0)Q}{(\alpha + \gamma K\beta L\tau_0) + SL}.$$

It follows from equations (4.2) and (4.7) that conductivity $\sigma_s = -e(u_p p_{ss} + u_n n_{ss})$ established in the sample under illumination consists of three terms:

$$\sigma_s = \sigma_0 + \Delta\sigma_{ph} + \sigma_m(\lambda), \qquad (4.8)$$

of which the first

$$\sigma_0 = e(u_p p_0 + u_n n_0) \qquad (4.9)$$

constitutes the contribution of equilibrium carriers to conductivity, the second

$$\Delta\sigma_{ph} = e K\beta L(u_n \tau_n + u_p \tau_p) \qquad (4.10)$$

is the contribution of nonequilibrium carriers to photoconductivity ($\Delta\sigma_{ph}$ is the photoresponse), the third term

$$\sigma_m(\lambda) = eu_mQ\,\frac{\gamma(n_0 + K\beta\tau_nL)}{\gamma(n_0 + K\beta\tau_nL) + SL} \tag{4.11}$$

is the contribution of metastable carriers to photoconductivity.

After the illumination is switched off and the concentration of nonequilibrium carriers becomes zero, $\sigma_m(\lambda)$ will tend to eu_mQ, while the dark conductivity σ_d will go to

$$\sigma_d = \sigma_0 + euQ.$$

The relaxation of conductivity in the dark follows the variation in carrier concentration. At first, a change of conductivity due to the recombination of nonequilibrium carriers takes place,

$$\frac{d\Delta n}{dt} = -\frac{\Delta n}{\tau_0} - \gamma\Delta n(Q - q), \tag{4.12}$$

followed by a slower change of conductivity due to the growth in the number of metastable carriers,

$$\frac{dq}{dt} = \gamma n_0(Q - q); \quad Q - q = (Q - q_{ss})e^{-\gamma n_0 t}. \tag{4.13}$$

In our approximation $(\tau_0 \ll \tau(\lambda))$, the relaxation of conductivity in the dark can be written in the form

$$\sigma = \sigma_0 + \Delta\sigma_{ph}e^{-t/\tau_0} + euQ - [euQ - \sigma_m(\lambda)]e^{-\gamma n_0 t}. \tag{4.14}$$

Equations (4.8)–(4.14) in fact determine the type of photoconductivity appearing in a semiconductor with s-centers when n_0 is not negligibly small and spontaneous capture of carriers by s-centers is significant. Note that in the steady state in the dark, all the s-centers are filled in the relevant temperature range, and the conductivity due to metastable carriers has its maximum value, so that when the semiconductor is illuminated, the concentration of metastable carriers will always decrease. In other words, the photoconductivity due to metastable carriers will be always negative in these conditions.

Various types of photoconductivity may arise depending on the relation of the parameters which enter equation (4.14).

1. The concentration of nonequilibrium carriers generated by light is less than the concentration of s-centers emptied by photons,

$$\frac{1}{eu_p}\,\Delta\sigma = K\beta L\tau_0 \ll Q - q_{ss}. \tag{4.15}$$

The steady-state conductivity under illumination is

$$\sigma_s \approx \sigma_0 + \sigma_m(\lambda) = \sigma_0 + eu_m\,\frac{(\gamma n_0 + \gamma K\beta L\tau_0)\,Q}{\gamma n_0 + SL + \gamma K\beta L\tau_0}, \tag{4.16}$$

and the steady-state value of dark conductivity is

$$\sigma_d = \sigma_0 + eu_m Q. \tag{4.17}$$

It follows from (4.16) and (4.17) that in this case the steady-state conductivity under illumination is less than the steady-state value of dark conductivity, i. e., the photoconductivity of the semiconductor with s-centers is negative in this case.

The photoresponse of negative photoconductivity is

$$\Delta\sigma_{\mathrm{ph}}^- = \sigma_s - \sigma_d = -eu_m Q\,\frac{SL}{\gamma n_0 + SL + \gamma K\beta L\tau_0}. \tag{4.18}$$

The contribution of positive photoconductivity $\Delta\sigma_{\mathrm{ph}}^+ = e(u_n\tau_n + u_p\tau_p)K\beta L$ is small compared with the negative component. This is the case which occurs when selenium is illuminated with blue light (relaxation curves in Figure 44, c and d).

2. The concentration of s-centers emptied by photons is less than the concentration of nonequilibrium carriers,

$$Q - q_{ss} < K\beta L\,(\tau_n + \tau_p). \tag{4.19}$$

In this case the photoresponse $\Delta\sigma_{\mathrm{ph}}$ is expressed by the relation

$$\Delta\sigma_{\mathrm{ph}} = e\,(u_n\tau_n + u_p\tau_p)\,K\beta L - eu_m Q\,\frac{SL}{\alpha + SL + \gamma K\beta L\tau_0} \tag{4.20}$$

and is less than the maximum change in conductivity $\Delta\sigma_{\max} \approx {}\approx eK\beta L\,(\tau_n u_n + \tau_p u_p)$, i. e., quenching of photoconductivity will occur (relaxation curves in Figure 44, a and b). Photoconductivity of this type is observed in activated amorphous selenium films when illuminated with light from the long-wave region of the visible spectrum.

3. Photons hardly empty any s-centers, the photoexcitation probability of a carrier from an s-center is small compared with the penetration probability into s-centers of free, thermally generated and photogenerated carriers:

$$\gamma n_0 + \gamma K\beta\tau_0 L \gg SL. \qquad (4.21)$$

In this case the photoconductivity component due to metastable carriers will have the same value, $eu_p Q$, both in the dark and under illumination. The photoconductivity is classical. When light is switched on, the photoconductivity grows to its steady-state value under illumination; when light is switched off, it falls to the dark level. The photoresponse and the relaxation curves will behave as if there were no s-centers in the semiconductor. Their presence will affect only the dark conductivity to which they contribute $eu_m Q$. As a result of this contribution, $\log \sigma_d$ is not a linear function of $1/kT$.

Finally, when $\alpha \to 0$, i.e., γn_0 is negligibly small, we have the case of anomalous photoconductivity.

In conclusion, note that in all the cases considered (1, 2, 3), when γn_0 is not negligibly small in a semiconductor with s-centers, the steady-state conductivity under illumination turns out to be dependent on light intensity, i.e., in all these cases we have normal photoconductivity (using the terminology of Chapter 1). When $\gamma n_0 + \gamma K\beta L\tau_0 \gg SL$ classical, normal photoconductivity is observed; when $K\beta\tau_0 L < (Q - q_{ss})$ the normal photoconductivity is negative. When $(Q - q_{ss}) < K\beta\tau_0 L$, self-quenching of photoconductivity occurs.

The foregoing discussion makes it possible to give a definite answer to the questions posed in the preceding section. When the temperature is raised, the anomalous photoconductivity, with the growth of γn_0, will turn into normal photoconductivity; this normal conductivity may be of the type observed in activated amorphous selenium films.

3. EXPERIMENTAL DATA LINKING THE NEGATIVE PHOTOCONDUCTIVITY IN ACTIVATED SELENIUM FILMS WITH S-CENTERS

In Sec. 1 of the present chapter it was suggested that the peculiar behavior of photoconductivity near the room temperature in activated amorphous selenium films was associated with the presence of s-centers in these films. In Sec. 2 it was shown that in a semiconductor containing such centers, different types of photo-conductivity may appear when the equilibrium carriers begin to play a substantial role in filling of the s-centers. We shall now present experimental data which confirm the suggestion that nega-tive photoconductivity in the short-wave region and self-quenching of photoconductivity in the long-wave region, observed in activated amorphous selenium films, are indeed due to photoexcitation of carriers from s-centers.

1. If negative photoconductivity is indeed due to the emptying of s-centers, the growth of conductivity from σ_s to σ_d after switching off the illumination is a process analogous to the growth in the dark of the steady-state anomalous photoconductivity. The latter is a process of thermal excitation of carriers into s-centers. There-fore, $d\sigma/dt$ (the rate of conductivity growth in the case of negative photoconductivity) and $d\sigma_a/dt$ should follow the same law, and in particular, should have the same temperature dependence.

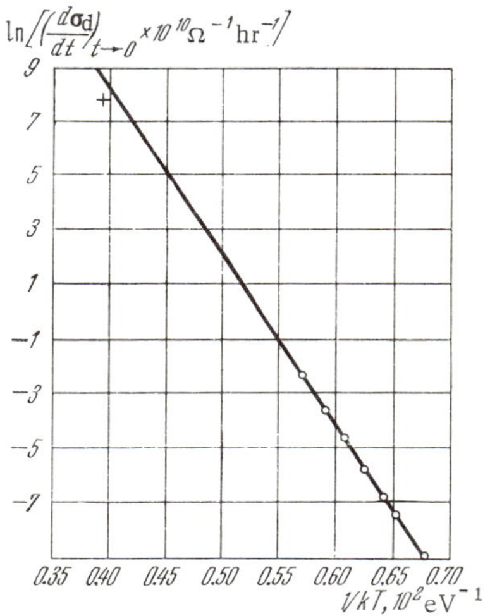

FIGURE 48. Temperature dependence of the rate of carrier excitation into long-life traps

To illustrate the extent to which this conclusion is true, the dependence of $d\sigma_a/dt$ on $1/kT$ is shown in Figure 48. Here, the ordin-ate gives $\ln(d\sigma_a/dt)$ and the abscissa gives $1/kT$ in units of $10^2\,\mathrm{eV}^{-1}$. The points represent the experimental values of the rate of growth of the dark anomalous photoconductivity. The straight line drawn through these points shows that $d\sigma_a/dt$ is an exponential function of temperature, with an activation energy $\Delta E = 0.60\,\mathrm{eV}$. The cross in Figure 48 represents the initial rate of growth of negative photo-conductivity at the time the illumination is switched off. This cross falls on the same line as the experimental points, which suggests that after switching off the illumination, at $T = 273°K$, the conductivity grows from the steady-state negative photoconductivity to the dark value at the same rate as it would change, at this temperature, on account of the filling of s-centers.

2. If negative photoconductivity indeed appears as a result of the emptying of s-centers by photons it will disappear after the blue-light illumination is switched off because of the thermal excitation of carriers into s-centers, which takes place at the rate $\alpha (Q - q)$ The photoconductivity will thus vary exponentially

$$\sigma_d - \sigma = (\sigma_d - \sigma_s) e^{-\alpha t} = \Delta\sigma_{ph}^- e^{-\alpha t}, \tag{4.22}$$

where $\Delta\sigma_{ph}^-$ is the photoresponse of negative photoconductivity. In reality, the relaxation of negative photoconductivity proceeds as shown in Figure 49, which shows the variation of the conductivity σ after the switching off of the light of wavelength $\lambda = 415\,nm$ (sample No. 9VK). The illumination is switched off at time $t = 0$. In Figure 50 the negative photoconductivity $\sigma_{ph}^- = \sigma_d - \sigma$ is plotted on a semilogarithmic scale ($\ln \sigma_{ph}^-$ vs. t).

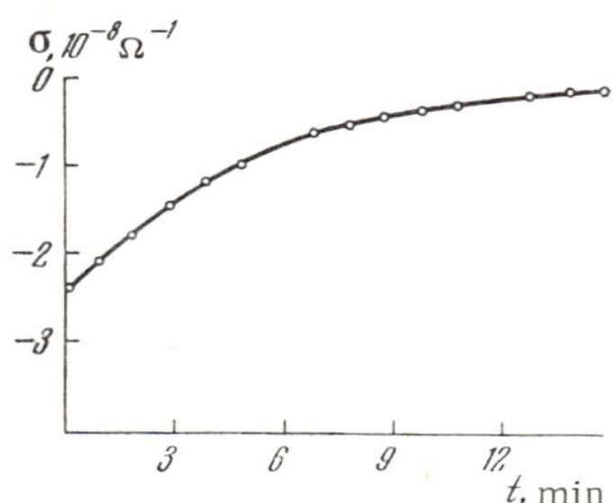

FIGURE 49. Relaxation curve of negative photoconductivity in a mercury-activated amorphous selenium film

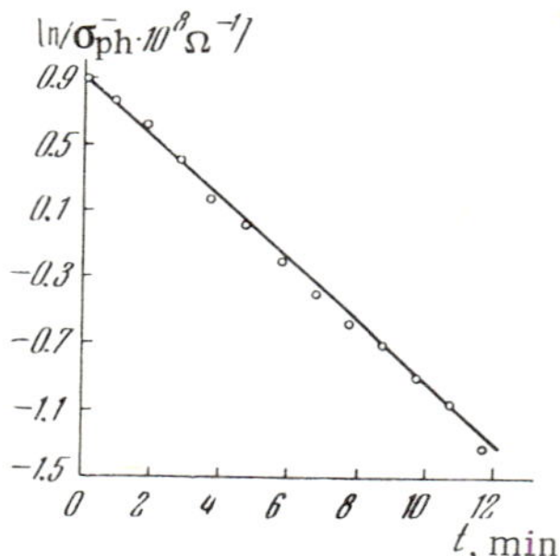

FIGURE 50. Relaxation curve of negative photoconductivity (sample No.9VK) on a semi-logarithmic scale

By (4.22)

$$\ln \sigma_{ph}^- = \ln \Delta\sigma_{ph}^- - \alpha t, \tag{4.23}$$

i. e., $\ln \sigma_{ph}^-$ is a linear function of time with the slope. It is seen on Figure 50 that the dependence of $\ln \sigma_{ph}^-$ on t is indeed linear and that the slope to the x-axis is $\alpha = \gamma n_0$.

3. The value of α determined from Figure 50 is $3 \cdot 10^{-3} \sec^{-1}$. However, α can be calculated from the values of the s-center parameters already established. Indeed, α constitutes the probability that one of the n_0 equilibrium carriers will penetrate into one of the s-centers in any given second. We shall denote by $D_{C'}$ the probability that an act by collision causes a carrier to penetrate into an s-center via the C' level. By N we shall denote the number of collisions with the s-centers which a carrier in the C'-band experiences. The product $D_{C'} N n_0$ thus represents the number of carriers penetrating into s-centers in one second, i. e.,

$$\frac{dq}{dt} = D_{C'} N n_0. \tag{4.24}$$

Further,

$$N = v/l = v \Delta S \, (Q - q), \tag{4.25}$$

where l is the free path of a current carrier in the process of collision with an s-center, ΔS is the effective cross-section of an s-center, v is the carrier velocity. Substituting this value of N in (4.24), we find

$$\alpha = v \Delta S n_0 D_{C'}. \tag{4.26}$$

The concentration of equilibrium carriers in the C'-band can be expressed by the relation

$$n_0 = (N_{C'} P_{V'})^{1/2} e^{-0.66/kT}.$$

Since the values of the effective density of states $(N_{C'}, P_{V'})$ in the bands and the exact value of α are not known, it is not possible to calculate α. We shall therefore confine ourselves to the estimation of its likely value. For this purpose, we have assumed $v = 6 \cdot 10^7$ cm/sec, $(N_{C'} \cdot P_{V'})^{1/2} = 10^{19}$. ΔS and $D_{C'}$ were calculated from the values of the s-center parameters: $\Delta S \simeq 30 \cdot 10^{-14}$ cm^2, $D_{C'} = 6.7 \cdot 10^{-6}$. Substituting all these numerical data in (4.26), we obtain $\alpha = 5.4 \cdot 10^{-3} \sec^{-1}$. This fit shows that the increase of conductivity in the dark from σ_s to σ_d may indeed be associated with the penetration of equilibrium carriers into s-centers via the C'-band.

4. It follows from (4.18) for the case when the positive photoconductivity is negligibly small that

$$\frac{e u_m Q}{\Delta \sigma_{\mathrm{ph}}^-} = \frac{K \beta L \tau_n \gamma + \alpha + SL}{SL} = A + B \frac{1}{L}, \tag{4.27}$$

where

$$A = \frac{K\beta\tau_n\gamma}{S} + 1; \quad B = \frac{\alpha}{S}.$$

(4.28)

By (4.27) the ratio $eu_mQ/\Delta\sigma_{ph}^-$ is a linear function of the inverse of light intensity. In order to clarify to what extent this corollary of the assumption that negative photoconductivity results from the emptying of s-centers is true, the dependence of $\Delta\sigma_{ph}^-$ on light intensity was studied for various wavelengths. Such a dependence is presented in Figure 51, based on measurement of negative photoconductivity at 0°C (sample No. 9VK) for illumination with monochromatic light of $\lambda = 415\,\mathrm{nm}$ ($\varepsilon = 3\,\mathrm{eV}$). In this figure the ordinate gives the ratio $\sigma_d/\Delta\sigma_{ph}^-$ and the abscissa $1/L$ in relative units.

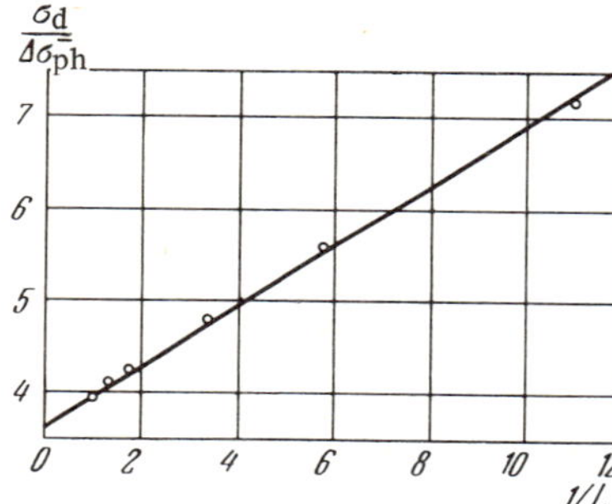

FIGURE 51. The photoresponse of negative photoconductivity as a function of light intensity

We see from Figure 51 that the dependence of $\sigma_s/\Delta\sigma_{ph}^-$ on $1/L$ is indeed linear. Similar linear functions are also obtained for light of wavelengths 427 nm ($\varepsilon = 2.9\,\mathrm{eV}$), 443 nm ($\varepsilon = 2.8\,\mathrm{eV}$), 460 nm ($\varepsilon = 2.7\,\mathrm{eV}$), 477 nm ($\varepsilon = 2.6\,\mathrm{eV}$), 495 nm ($\varepsilon = 2.5\,\mathrm{eV}$), 515 nm ($\varepsilon = 2.4\,\mathrm{eV}$), 540 nm ($\varepsilon = 2.4\,\mathrm{eV}$), 565 nm ($\varepsilon = 2.3\,\mathrm{eV}$). Under the action of all these wavelengths, negative photoconductivity appears in the films, and its magnitude varies with light intensity in a way which indicates that negative photoconductivity is indeed due to the emptying of s-centers.

From the dependence of $\sigma_d/\Delta\sigma_{ph}^-$ on $1/L$, it is possible, making use of (4.27) and (4.28), to determine quantities proportional to $\left(\frac{\alpha}{S}\right)$ and $\left(\frac{K\beta\tau_0\gamma}{S} + 1\right)$ which depend on the s-center parameters. The quantities were previously determined for a wide spectral range from data of anomalous photoconductivity measurements (Secs. 4, 7, Chapter 3). It follows from (4.28) (if negative photoconductivity indeed results from the emptying by photons of s-centers) that $1/B$ is proportional to the cross-section S. To prove this, Figure 52 plots on a logarithmic scale the dependence of the cross-section S on the photon energy. On this figure, the ordinate represents the excitation cross-section S, and the abscissa the corresponding photon energy $h\nu$. The curve plots values of S previously determined from the relaxation curves of anomalous photoconductivity. The points represent values of S calculated from negative photoconductivity data at $T = 0°C$. For both values, the cross-section for photons of energy $\varepsilon = 3.0\,eV$ $(\lambda = 415\,nm)$ is taken as unity. It is seen that the two series of measurements give consistent results.

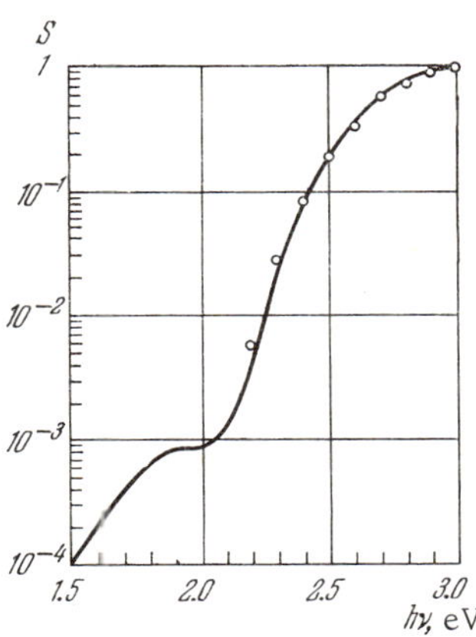

FIGURE 52. The excitation cross-section of a carrier from an s-center according to anomalous photoconductivity data (solid curve) and negative photoconductivity data (points)

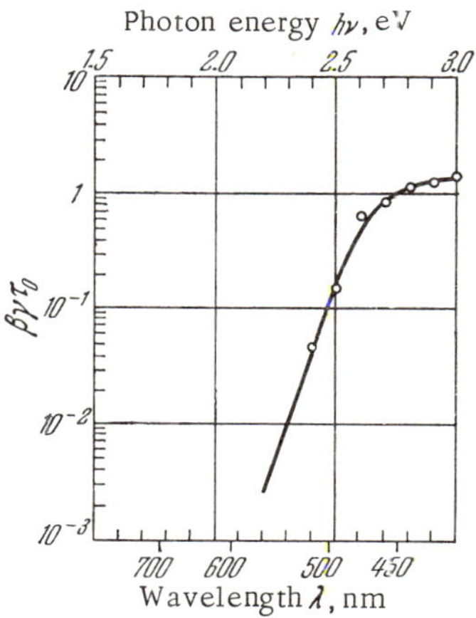

FIGURE 53. Values of $\beta\gamma\tau$, according to anomalous photoconductivity data at $T = 123°K$ (solid curve) and negative photoconductivity data at $T = 273°K$ (points) in activated selenium films (sample No.9VK)

5. Let us compare the values of $\beta\gamma\tau_0$ calculated from negative photoconductivity data (from values of the coefficient A) with those from the relaxation curves and anomalous photoconductivity data. In the latter case, it is not $\beta\gamma\tau_0$ that is calculated but the proportional quantity $\sigma_a S/K = e \cdot u_m Q\beta\gamma\tau_0$. Figure 53 shows the dependence of $\beta\gamma\tau_0$ on the photon energy $\varepsilon = h\nu$. The ordinate represents on a logarithmic scale $\beta\gamma\tau_0$ in relative units, and the abscissa the photon energy in eV and the corresponding wavelength. The solid curve is constructed from measurements of anomalous photoconductivity $\sigma_a S/K$ and the points represent values of $\beta\gamma\tau_0$ calculated from negative photoconductivity measurements. It is seen on Figure 53 that the two series of measurements fit well.

The data considered in the present section show that the regular features observed in mercury-activated amorphous selenium films, i.e., the conductivity relaxation kinetics after switching off the blue-light illumination, the rate of change of conductivity, the dependence of the steady-state value of negative photoconductivity on light intensity and on wavelength, are indeed such, both qualitatively and quantitatively, as expected if negative photoconductivity were due to the emptying of s-centers.

We may therefore assert that all the specific features of photoconductivity observable in mercury-activated amorphous selenium films, i.e., anomalous photoconductivity at low temperature, negative photoconductivity at room temperature when the film is illuminated with 400—500 nm light, self-quenching of photoconductivity when these films are illuminated with 600—800 nm light, are due to the existence in these films of s-centers.

An indirect consequence of the fit of the curve $\beta\gamma\tau_0$ constructed in Figure 53 from negative photoconductivity data to the points $\beta\gamma\tau_0$ obtained from anomalous photoconductivity data is that the mobility u_m in films illuminated with light of various wavelengths is independent of the wavelength within the spectral region 400—530 nm.

4. DETERMINATION OF THE DEGREE OF FILLING OF S-CENTERS IN ANOMALOUSLY PHOTOCONDUCTIVE AMORPHOUS SELENIUM FILMS

A comparison of anomalous photoconductivity data in amorphous selenium films at low temperature with the negative photoconductivity data at room temperature ($T = 293°K$) in the same films makes it possible to determine the degree of filling (occupation) of s-centers by current carriers when the films are in the anomalous photoconductivity state.

At room temperature in the dark, all s-centers are filled. The contribution eu_mQ of metastable carriers to the dark conductivity at this temperature is quite significant; however, the main role in dark conductivity is played by the equilibrium carriers. When the sample is cooled, the contribution of these components changes substantially. The conductivity due to equilibrium carriers falls exponentially as the temperature is lowered, with an activation energy of 0.16 eV. When the temperature is reduced from $T = 293°K$ to $T = 123°K$, this component of dark conductivity decreases by a factor of 6,600. For such a temperature change, the dark-conductivity component due to metastable carriers is altered substantially less. When the sample is cooled in the dark, the concentration of metastable carriers remains unchanged, and their mobility does not vary exponentially but according to a power law. Therefore, at $T = 123°K$, the component due to the metastable carriers makes the predominant contribution to the dark conductivity.

TABLE 6

No. of sample	$\eta(\lambda)$			No. of sample	$\eta(\lambda)$		
	$\lambda = 415\,nm$	$\lambda = 730\,nm$	$\lambda = 886\,nm$		$\lambda = 415\,nm$	$\lambda = 730\,nm$	$\lambda = 886\,nm$
6VK	0.06	0.125	0.21	11VK	0.042	0.27	0.47
7VK	0.032	0.14	0.16	15VK	0.045	0.37	0.43
8VK	0.037	0.13	0.18				

The total dark conductivity σ_d can be separated into the components $\sigma_m = eu_m Q$ and σ_0 by analyzing the cooling and heating curves of the sample in the dark and after illumination with white light. In Sec. 6, Chapter 5 this problem will be considered in more detail. Isolating the component σ_m from σ_d at $123°K$ and comparing it with the value of anomalous photoconductivity, we can apply the relation

$$\sigma_m(\lambda) = \eta(\lambda) \cdot eu_m Q \tag{4.29}$$

to determine the factor $\eta(\lambda)$ which characterizes the degree of filling of s-centers with carriers following illumination of the sample at a given wavelength λ.

Values of $\eta(\lambda)$ calculated from (4.29) for various samples are presented in Table 6.

5. THE NATURE OF THE NONANOMALOUS PHOTOCURRENT COMPONENT OBSERVED IN AMORPHOUS SELENIUM FILMS AT 120–200°K

It has been shown that in anomalously photoconductive selenium films some contribution to conductivity is frequently observed at $T = 123°K$ from a normal component, whose steady-state valve $\Delta\sigma_n$ depends on light intensity. It is because of the existence of this photocurrent component that the conductivity changes by $\Delta\sigma_n$ after the illumination is switched off, from σ_s to $\sigma_d = \sigma_m + \sigma_0$. The normal photocurrent component in anomalously photoconductive films can be isolated by repeated illumination of the sample with light of a given spectral composition.

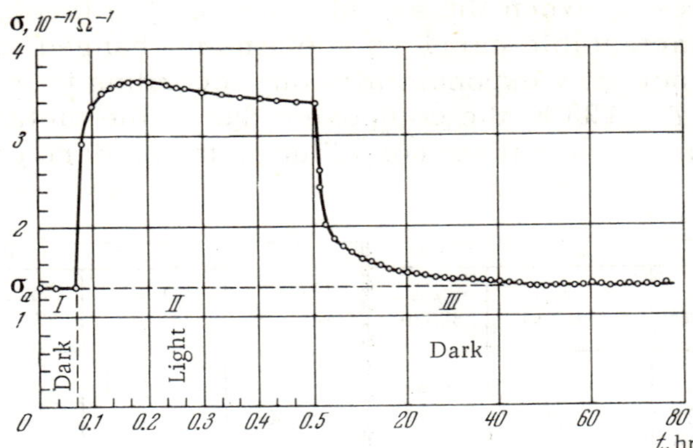

FIGURE 54. Relaxation curve of the nonanomalous component of photoconductivity ($T = 123°$ K) under repeated illumination of the sample at $\lambda = 730$ nm

Since $\sigma = \sigma_n + \sigma_a$ and since the anomalous component of photoconductivity σ_a remains unchanged both in the dark after the illumination is switched off and under repeated illumination of the sample with light of a given spectral composition, it is clear that the entire change in σ observed under repeated illumination is due to a change in σ_n. Consequently, the relaxation curves of σ under repeated illumination with light of a given composition are the relaxation curves of σ_n.

Figure 54 shows a conductivity relaxation curve for repeated illumination of an activated selenium film at $\lambda = 730$ nm. There are three regions on this figure. Region I presents part of the photoconductivity relaxation curve after switching off the illumination of a given wavelength, when the steady-state value of conductivity has been established. Light of $\lambda = 730$ nm is then switched on again. The resulting conductivity relaxation is shown by region II. Self-quenching of the nonanomalous component is distinctly seen. Finally, region III presents the relaxation curve after switching off the light. In this region the time scale (abscissa) has been greatly shortened.

Figure 55 shows a conductivity relaxation curve observed when the sample is illuminated with light of wavelength $\lambda = 420$ nm. There are two regions on this figure: region I which corresponds to the steady-state value of conductivity under illumination, and region II which shows the conductivity of the sample after switching off the illumination. When the sample is illuminated at $\lambda = 420$ nm, the conductivity under illumination is less than its dark value.

After the illumination is switched off, the conductivity increases. This means that the nonanomalous component of photoconductivity increases and that the nonanomalous component of photoconductivity is negative. Thus the nonanomalous component of photoconductivity observable at low temperatures exhibits some features of normal photoconductivity of activated selenium films at room temperature, namely, negative photoconductivity for short-wave light and self-quenching of photoconductivity for long-wave light. It can thus be expected that this component of the photocurrent (or a substantial part of it) is associated with centers of s-center type, differing from the usual s-centers in that equilibrium carriers fill them in large numbers already at 123°K.

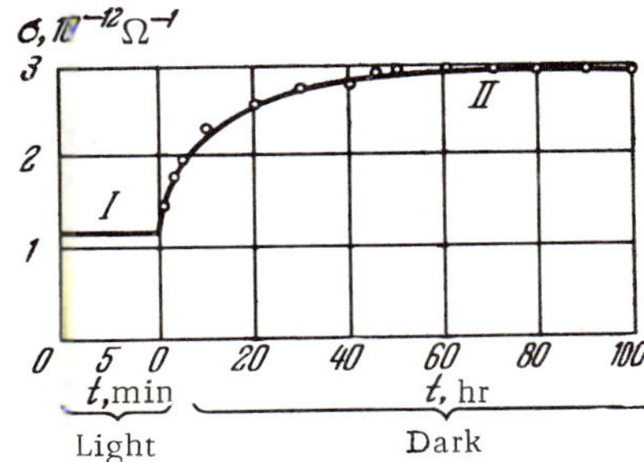

FIGURE 55. Relaxation curve of nonanomalous photoconductivity after switching off the illumination ($\lambda = 420$ nm)

We shall subsequently identify these centers as s-centers. The penetration rate in the dark of carriers into s-centers at $T = 123°$K is millions of times larger than the penetration rate of carriers into s-centers at the same temperature. Since the normal component of anomalously photoconductive films contains information about centers similar to s-centers, it is of interest to study their properties. Unfortunately, the properties of such centers have so far been little studied. Some available data will be presented below.

Figure 56a shows the spectral distribution of the photoresponse of the normal component for sample No. 1VK in the wavelength range 400—900 nm, at $T = 123°$K. For comparison, Figure 56b presents the spectral distribution curve of anomalous photoconductivity in the same spectral range (in relative units). The similarity of the two spectral distributions, which possibly reflects a similarity in the nature of the two types of centers, is remarkable.

The spectral distribution of the anomalous component shown on Figure 56a, together with the relaxation curves of Figures 54 and 55, lead to the conclusion that this component in its turn is made up of two parts, a positive photoconductivity $\Delta\sigma^+$ and a negative photoconductivity $\Delta\sigma^-$:

$$\Delta\sigma_n = \Delta\sigma^+ - \Delta\sigma^-.$$

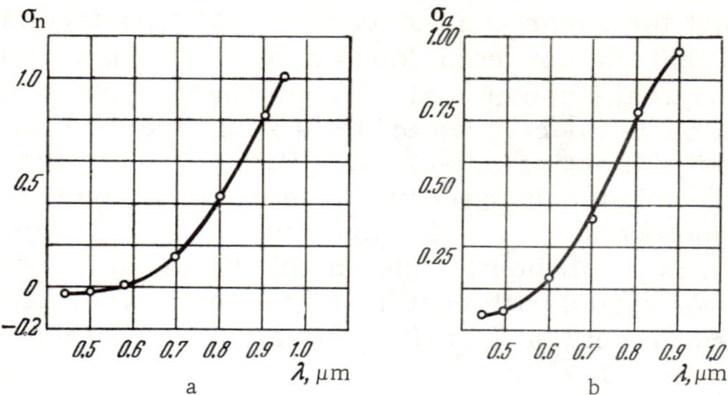

FIGURE 56.   Spectral distribution of the nonanomalous (a) and the anomalous (b) component of photoconductivity

In the long-wave region of the spectrum $\Delta\sigma^+$ is relatively large, and in the visible spectrum it decreases with decreasing wavelength. $\Delta\sigma^-$, on the other hand, decreases with increasing wavelength.

From data of the preceding section and from the shape of the relaxation curve (Figure 55), we conclude that the concentration Q' of s-centers is small compared with the concentration Q of s-centers in the sample.   Indeed, it follows from the relaxation curve (Figure 55) that $eu_m Q'$ constitutes a substantial contribution to $\sigma_d(420)$, for

$$\Delta\sigma^- = eu_m \left(Q' - q'_{ss}\right) \simeq {}^2/_3\, \sigma_d.$$

But even if it is assumed that the whole value of $\sigma_d(420)$ is due to s'-centers, $eu_m Q'$ will still be small compared with $\sigma_d(730) = eq_{ss}(730)u_p$.   The ratio $\sigma_d(730)/\sigma_d(420) \simeq 10$.   Further, it is shown in the preceding section that q_{ss} is small compared with Q (being approximately $10^{-1} Q$).   Consequently, the ratio Q'/Q does not exceed 10^{-2}.

The activity of such a small amount of s'-centers may become noticeable against a background of nearly 100 times as many s-centers because at 123°K the s'-centers are filled with carriers to a considerable degree (in the dark, almost completely), whereas the s-centers are almost empty (at $\lambda = 420$ nm).

The dependence of $\Delta\sigma_n$ on light intensity is shown on Figure 57, a and b.   On Figure 57a the ordinate gives $\Delta\sigma_n$ in relative units and the abscissa gives the light intensity L, also in relative units.   On Figure 57b the abscissa gives $\sqrt{L}$ and the ordinate $\Delta\sigma_n$.   It is seen that $\Delta\sigma_n \sim \sqrt{L}$.

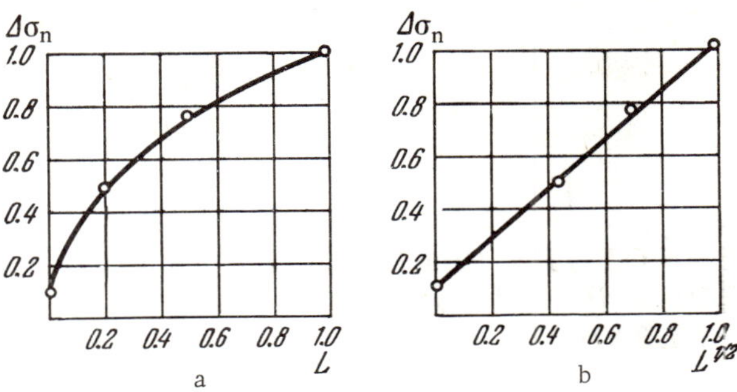

FIGURE 57. The photoresponse of the nonanomalous component of photoconductivity in anomalously photoconducting amorphous selenium vs. light intensity

The analysis of this chapter shows the following:

1. In a semiconductor with s-centers, anomalous photoconductivity may be accompanied by other phenomena. For intensive capture of equilibrium carriers by s-centers and the case of s-centers which are filled in the dark, the existence of s-centers in the semiconductor will produce either negative photoconductivity (if the photoresponse of the normal component $\Delta\sigma_n$ is small and $\Delta\sigma_{ph}^-$ is large), or self-quenching of photoconductivity (if $\Delta\sigma_{ph}$ is a large value and $\Delta\sigma_{ph}^-$ is small).

2. In mercury-activated amorphous selenium films, the photoconductivity at room temperature behaves precisely in this manner. Negative photoconductivity is observed in blue light, self-quenching in red. Both negative photoconductivity in blue light and self-quenching in red are due to the emptying of s-centers by photons.

3. Anomalous photoconductvity has so far been observed only in activated selenium films. Negative photoconductivity is encountered more frequently. The negative photoconductivity observed in a number of cases /60—66/ (in cuprite /60/, in silver-activated PbS films /61/, in cuprous oxide when it contains colloidal particles) is possibly due to centers similar to s-centers.

4. In addition to s-centers, selenium contains a small quantity of s'-centers, which in the dark fill up with equilibrium carriers at a marked rate also at temperatures below 200°K. The concentration of s'-centers determines the magnitude of the nonanomalous component of photoconductivity.

Chapter 5

SOME PROBLEMS OF THE PREPARATION TECHNOLOGY OF ANOMALOUSLY PHOTO-CONDUCTIVE SELENIUM

1. SOME PROPERTIES OF AMORPHOUS SELENIUM

Selenium is found in various allotropic forms. The amorphous form can be obtained either by supercooling liquid selenium or by condensation in vacuum. The resistivity of amorphous selenium at room temperature is about $10^{12}\,\Omega \cdot cm$. From measurements of transit time of drifted injected carriers in thin amorphous selenium films, the effective mobility of electrons was found to be $(4.7-5.5)10^{-3}\,cm^2/sec \cdot V$ and of holes $0.13-0.15\,cm^2/sec \cdot V$. The mobility varies exponentially with temperature with an activation energy of $0.25\,eV$ for electrons and $0.14\,eV$ for holes /17/.

Amorphous selenium has strong absorption in the ultraviolet and in the short end of the visible spectrum. It is almost transparent to red light. The spectral distribution of the absorption coefficient of selenium not activated by mercury is shown on Figure 36, by the curve $K = f(\varepsilon)$.

Amorphous selenium is photoconductive with a red boundary of $\simeq 650\,nm$ /30, 12/. The red boundary of the photoeffect does not coincide with the absorption band edge.

The width of the energy gap in amorphous selenium films, determined by various methods, has somewhat differing values. From data on the measurement of the reflectivity of selenium /82/, the value of E_g is found to be $\simeq 2.2-2.3\,eV$, from optical absorption data $\simeq 2.3\,eV$ /83/, from photoconductivity data $\simeq 2.5\,eV$ /28/.

A detailed diagram of energy states in amorphous selenium is presented on Figure 7. Its characteristic feature is the presence of exciton levels distant $0.08\,eV$ from the bottom of the conduction band. The existence of such levels was discovered through the investigation of the fine structure of the absorption band edge of hexagonal selenium /84/. No fine structure was discovered in the absorption band edge of amorphous selenium, a fact due to the

existence of exciton states at the gap edges.   The existence of
exciton levels leads to the noncoincidence of the photoeffect boundary
and the absorption edge.

The deep electron states near the Fermi level (Figure 7), whose
density falls exponentially, were determined from measurements
of space-charge limited currents /85/.   The photoconducting transitions from these states, situated below the Fermi level, are very
weak, since the quantum yield for this process for photon energies
of 1.2—1.85 eV is negligibly small /11/.

The identification of the sign of majority carriers in amorphous
selenium is a complicated problem.   Amorphous selenium is a
thermodynamically unstable system undergoing crystallization at
room temperature.   The crystalline phase embryos may considerably affect the electrical properties of amorphous selenium.

The presence of oxygen in selenium has a significant effect.
Oxygen-free amorphous selenium or selenium with compensating
Cd and Mn impurity has electron-type conductivity /68/.   In oxygen-
containing selenium hole conductivity predominates /85/.

Summaries of data on the electrical properties of selenium
including its properties in the film form are given in /15, 67/.

2. PURITY REQUIREMENTS FOR SOURCE SELENIUM AND SUBSTRATE MATERIAL

When selenium is condensed in a vacuum of the order of 10^{-5}—
10^{-4} mm Hg, the amorphous selenium layer may appear in one of
two forms.   The first form exists up to 20°C, the second up to 60°C.
The transition from one amorphous form to another occurs without
the appearance of a crystalline phase.

Amorphous selenium contains small regions of hexagonal structure consisting of chains of a few hundred atoms arranged in spirals.
The ends of such chains act as traps for current carriers.   The
conductivity along the chains is relatively high.   A carrier drifts
along the chain and is captured at the site where it breaks off.   In
the case of sublimation on a cold substrate, the formation of an
amorphous phase with some monoclinic impurity is observed.

In evaporated amorphous selenium films used to obtain anomalous
photoconductivity, we did not succeed in establishing the existence
of a crystalline phase in a quantity sufficient for its detection by
X-ray methods.

The amorphous selenium layers in which anomalous photoconductivity was investigated were prepared by vacuum condensation

of selenium vapor on a dielectric substrate. Layers obtained by evaporation of selenium of various purities were investigated, and also selenium specially purified from mercury impurities by a method developed in the Institute of Nuclear Physics semiconductor laboratory of the Kazakh Academy of Sciences /69—70/. Selenium containing mercury in quantities of up to 10^{-2}% was also evaporated. In all cases, after appropriate treatment with mercury and exposure to light, the samples acquired anomalously photoconducting properties at a certain temperature. Note that the introduction of up to 10^{-2}% mercury into the selenium used for the preparation of vacuum-deposited layers did not change the anomalously photoconducting properties of the samples to any noticeable extent (no experiments were carried out with selenium containing over 0.2% mercury). To impart anomalous photoconductivity to these deposited layers, it was necessary to carry out the same treatment in mercury vapor as for layers obtained by the evaporation of selenium containing less than 10^{-5}% mercury impurities.

The experiments show that the purity of the source selenium is not critical for the preparation of anomalously photoconductive amorphous selenium films.

No substantial influence of the substrate material was observed on the anomalously photoconductive properties of amorphous selenium films. Glass, quartz, Plexiglas, vinyl plastics were completely equivalent in this respect. After appropriate treatment with mercury vapor and light, the films acquired anomalous photoconductivity. The insensitivity of anomalous photoconductivity to the substrate material (it is only necessary that the substrate resistivity be sufficiently high) proves that the selenium-substrate boundary does not play any essential role in the formation of anomalous photoconductivity. This fits the conclusion reached in Sec. 1, Chapter 3 that the s-centers responsible for the appearance of anomalous photoconductivity are situated in a surface layer of the selenium film (in the valleys of its surface relief).

However, to acquire anomalously photoconductive properties, the selenium layer has to be cooled to 100—130°K, and then be repeatedly heated and cooled in the process of its exploitation. The substrate material is therefore of the greatest practical importance in the preparation of thermally stable samples.

Best for this purpose are substrates of insulating materials whose linear expansion coefficient is close to that of selenium. Plexiglas substrates are preferable in this respect to glass or quartz ones. Indeed, the linear thermal expansion coefficient of Plexiglas is $(50-60) \cdot 10^{-6}$ deg^{-1}, whereas that of glass is $8 \cdot 10^{-6}$ deg^{-1} and of amorphous selenium $40 \cdot 10^{-6}$ deg^{-1}. Moreover it is much simpler to attach the leads to Plexiglas than to glass or quartz.

Because of the great plasticity of Plexiglas, mechanical binding of copper wires to the substrate is more reliable than with glass. When the temperature is lowered, the favorable relation of the linear expansion coefficients of Plexiglas and copper (α for Plexiglas is larger than the α for copper) improves the wire-to-substrate band, whereas for glass and quartz, whose expansion coefficients are less than that of copper, the wire-to-substrate band deteriorates.

Amorphous selenium layers on Plexiglas substrates endured repeated temperature changes from 290–300°K to 100–120°K. On many samples, experiments were conducted continuously for several months.

Gold layers deposited on the substrate usually acted as electrodes. However, the electrode material apparently does not have any substantial effect on the anomalously photoconductive properties of the samples. Anomalous photoconductivity was observed in selenium layers with electrodes from the following metals: silver, indium, aluminum, copper, zinc, cadmium, tellurium, nickel, platinum. These materials include those with a work function both higher and lower than that of selenium. The work functions of the electrode materials /11/ are presented in Table 7.

TABLE 7. Work functions of electrodes

Material	Se	Au	Ag	In	Al	Cu	Zn	Cd	Te	Ni	Pt
Work function, eV	4.72	4.25	4.31	4.09	4.25	4.4	4.24	4.1	5.0	4.5	5.3

It follows from Table 7 that anomalous photoconductivity is observed in mercury-activated selenium films with electrode work function both higher and lower than that of selenium. Anomalous photoconductivity is thus not associated with barrier effects at the electrodes.

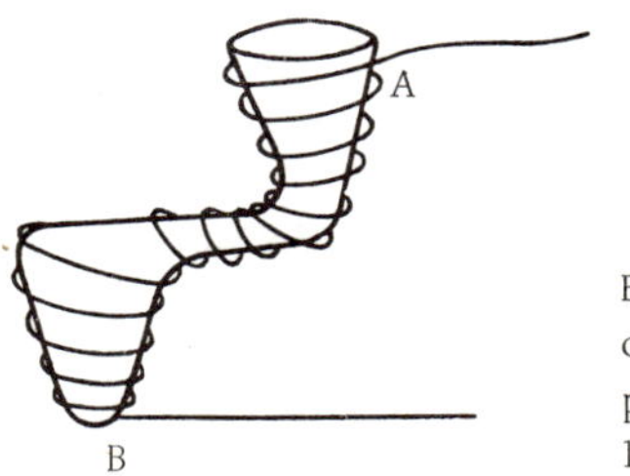

FIGURE 58. Schematic diagram of the evaporator used in the preparation of amorphous selenium layers

To prepare quality selenium layers free from through-pore defects, the construction of the evaporator is very important. In order to avoid the appearance of through pores in selenium layers, it is necessary to take care that no selenium "bits" mix up with the molecular stream. For this purpose, the evaporator has a special geometry: selenium molecules can leave the evaporator only after colliding a number of times with the walls, which are kept at a higher temperature than the evaporating selenium. As a result, the selenium vapor leaving the evaporator is close to molecular in its composition.

The schematic diagram in Figure 58 shows the basic evaporator, which is a bent quartz test-tube with a nichrome wire spiral, wound in such a way that the temperature of the wall A is higher than the temperature in the region B, where the source selenium is placed in the test-tube.

3. TREATMENT OF AMORPHOUS SELENIUM FILMS WITH MERCURY VAPOR

Treatment with mercury plays a decisive role in imparting anomalously photoconducting properties to condensed amorphous selenium layers /6, 10, 28/. For this purpose, the substrate with the deposited selenium layer (the sample) is placed in a special container. A schematic diagram of the apparatus for the treatment of selenium layers with mercury vapor is shown on Figure 59. The container 1, which contains mercury 5, is sealed by means of a packing ring 3 and lid 2 to which the sample 4 to be treated is attached. The leads 6 are inserted through the lid. The mercury treatment is monitored by measuring the resistance of the sample, which changes in the course of the treatment. The magnitude of the resistance is recorded by a EPPV-60 potentiometer. The resistance curves (Figure 60) change depending on whether the treatment was carried out in the dark (curve I) or under illumination (curve II). Four characteristic sections can be distinguished on these curves /57, 72/. Section 1 is the region of low conductivity. The current through the sample is small, lying at the sensitivity threshold of the instrument. Section 2 is the first region of fast conductivity growth, section 3 is a region of constant or slowly growing conductivity, section 4 is the second region of fast conductivity growth. The observed curves are satisfactorily explained on the basis of the assumption that a high-conductivity (relative to selenium) sublayer is formed at the selenium layer surface.

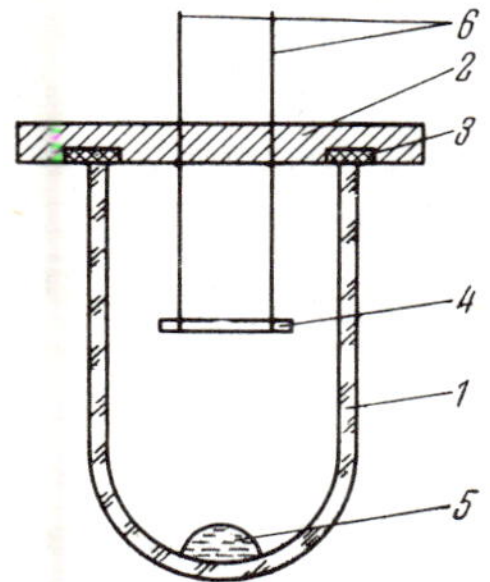

FIGURE 59. Schematic diagram of the setup for mercury-vapor treatment of amorphous selenium layers

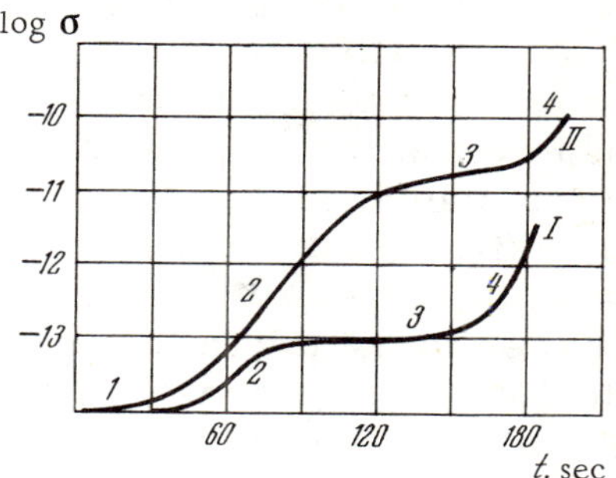

FIGURE 60. Time curves of the conductivity of selenium layers in mercury vapor

At the beginning of the treatment, when the conducting sublayer has not yet formed, the conductance of the sample is determined by the conductivity of the selenium layer. For sample thickness of the order of $1\,\mu$m and interelectrode distance of a few millimeters, the conductivity of the sample does not exceed $10^{-13}\,\Omega^{-1}$, and the current through the sample is far below the sensitivity threshold of the instrument. During stage 1 of the treatment process, a small number of isolated high-conductivity islands are formed, but they remain disjoint and therefore have little effect on the conductivity of the whole sample.

The fast growth of conductivity during the second stage of the treatment may be due to the merging of the isolated islands, formation of conducting bridges, or appearance of conductivity of the Borzjak-Hartmann type /73, 74/. The fast growth of the sample conductivity evidently persists until the resistance of the conducting film R_f becomes small compared with R_{Se}, the resistance of the selenium layer above the hidden electrode. The growth of conductivity is then slowed down or stops altogether (region 3). Since the resistivity of selenium decreases under illumination, sample conductivity will grow under illumination to much higher values than in the dark. The growth phase will therefore last longer, as is clear from the curves in Figure 60.

In region 3, $R_f \ll R_{Se}$, i.e., the conducting sublayer formed on the selenium surface serves as a second electrode which lies above the submerged electrode.

The resumed growth of conductivity (region 4 on the conductivity curve), which is accompanied by a sharp decrease of photoresponse and the appearance of instabilities in the resistance of the sample, is due to the formation of mercury bridges across the selenium layer and the resulting short-circuiting of the photoactive layer.

This may be due to the presence of microholes (channels) which facilitate the diffusion of mercury.

In some samples (usually the thicker ones), region 4 is not reached even after treatment of the selenium layer with mercury vapor for many hours. The resistance of the conducting film drops to a few kilohm, although the resistance of the entire sample remains quite large ($10^{-7}-10^{9}\,\Omega$).

In order to establish the extent of the mercury-vapor treatment of selenium, let us choose a parameter measuring the quality of the anomalously photoconductive sample. Since spectral memory is the basic characteristic of anomalous photoconductivity, it is clear that the best sample is that for which the difference in the photocurrent between the two ends of the visible spectrum ($\lambda_0 = 420\,$nm and $\lambda_1 = 740\,$nm) is the largest.

However, the magnitude of the photocurrent depends not only on the carrier concentration but also on the conditions of carrier transport, in particular on carrier mobility. To eliminate the effect of mobility, we use the photocurrent ratio at the relevant wavelengths

$$Z = \frac{i\,(\lambda_1)}{i\,(\lambda_0)}$$

as a measure of the quality of the anomalously photoconductive sample. The ratio Z thus defined does not depend on the potential across the sample /75/. We shall call Z the color response of the anomalous photoconductor. For illustration, Figure 61 shows the current-voltage characteristics of mercury-activated films. Curve I is the current-voltage characteristic of a sample after exposure at $\lambda = 740\,$nm, and curve II is the current-voltage characteristic after exposure at $\lambda = 420\,$nm. All the current values shown in Figure 61 are given in relative units. The current which passes through the sample after illumination at $\lambda = 740\,$nm with a voltage of 20 V across the sample is chosen as unity. The ratio $i\,(740)/i\,(420) = Z$ as a function of the voltage V across the sample is presented in Figure 62. It is seen that this ratio indeed does not depend on the voltage V.

To find the optimal degree of mercury-vapor treatment, the variation of Z in the process of treatment was investigated. In order to bring the sample into the anomalously photoconductive state, it is necessary to cool it and then submit it to a prolonged treatment with light. It is therefore practically impossible to contiuously monitor the value of Z during mercury treatment of the sample. A stepped treatment of the sample was thus adopted, replacing the continuous process /75/. The sample was immersed in mercury vapor for a fairly short time, until its resistance fell to a certain value. The sample was then transferred to a cryostat, cooled and submitted to illumination with white light for a time necessary to impart anomalously photoconductive properties to the sample. The spectrum of anomalous photoconductivity was then investigated and the ratio Z determined. After the determination of Z, the sample was again submitted to a repeated mercury-vapor treatment for a short time, until its resistance decreased to another control value. The sample was then again cooled and treated with light until the appearance of anomalously photoconductive properties, and the ratio Z was again determined. This procedure was repeated several times.

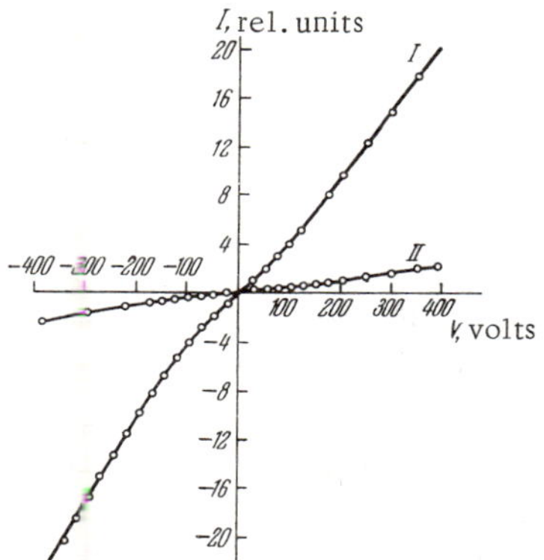

FIGURE 61. Current-voltage characteristics of anomalously photoconductive selenium

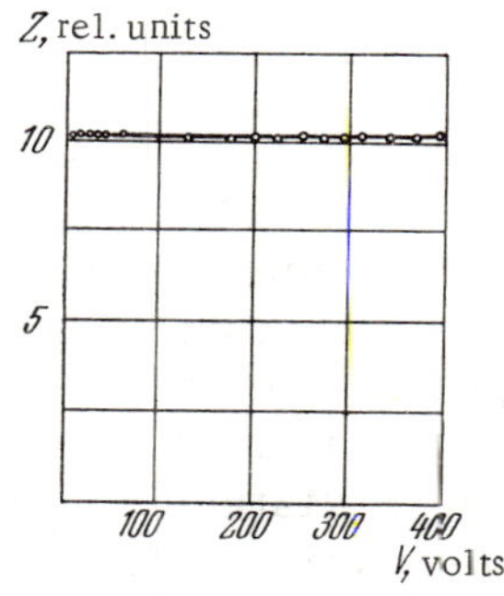

FIGURE 62. The ratio z vs. the voltage across the sample

As a result of such experiments it became clear that there exists some optimal degree of mercury-vapor treatment, characterized by the resistance R_{opt}, for which the color response Z is maximum. Z is found to be less than the optimal both for insufficient treatment with mercury vapor, when the sample resistance R is higher than optimal, and for excess treatment $(R < R_{opt})$.

When the deviation from the optimal condition is large, $R \ll R_{opt}$, Z tends to unity, i.e., the sample loses its spectral memory.

Figure 63 plots the dependence of Z on log R for one of the samples investigated. It is distinctly seen that the range of R values corresponding to the optimal value of Z is quite narrow. For the sample investigated in /75/ (Figure 63), R_{opt} is between $10^8 - 10^9\,\Omega$.

This range of R depends on the conditions of preparation of the sample, namely on the conditions of deposition of selenium, on the electrode geometry, and on the material of the electrodes. For fixed conditions of sample preparation, the value of R_{opt} varies but little from sample to sample, so that it becomes possible to determine R_{opt} for one of the samples of a given series and then apply the same criterion for all the samples of the series.

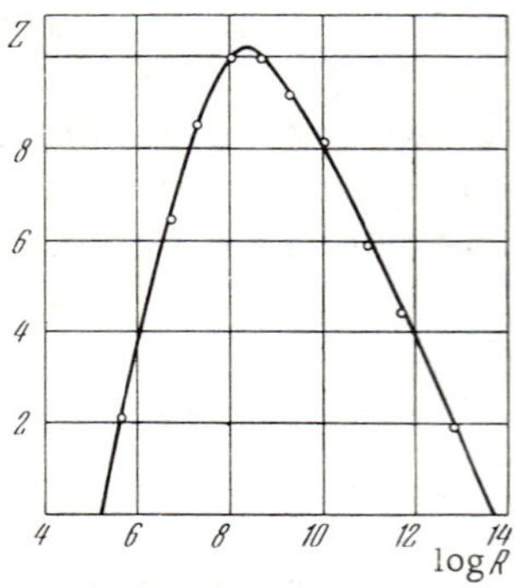

FIGURE 63. The dependence of the anomalously photoconductive properties of a sample on the degree of mercury-vapor treatment

The dependence of Z on R presented in Figure 63 can be explained on the basis of the equivalent circuit of the sample. At the beginning of treatment with mercury vapor, when R is large, the photoactive section is connected to the measuring circuit in series with a large resistance. The change in the conductivity of the photoactive section affects comparatively little the resistance of the whole sample. Z is therefore close to unity. As the mercury-vapor treatment of the sample progresses, the conductivity of the conducting sublayer increases, the resistance of the selenium layer between the electrodes markedly decreases, and the photoactive section begins to make the major contribution to the resistance. Z becomes quite large. Under further mercury-vapor treatment of the sample, the resistance of the sublayer falls to such an extent that it starts to shunt the photoactive section, and its contribution to the overall resistance becomes small. In this case $Z = 1$.

4. THE EFFECT OF THE DEGREE OF MERCURY ACTIVATION OF A SELENIUM FILM ON THE PARAMETERS OF S-CENTERS

In order to clarify the interaction mechanism of mercury and selenium, the s-center parameters were investigated, i. e., the cross-section S and the quantum yield β' of a given sample at

several stages of mercury activation.   For this purpose, the mercury
treatment was discontinued when the sample resistance had decreased to the given control values $R_1 = 10^{11} \Omega$.   The film was
then cooled to 123°K and formed with light; the cross-section S
and the steady-state anomalous photoconductivity were determined
for the whole spectral range from 400 to 1,300 nm by the method
mentioned earlier (Sec. 3, Chapter 3).

The sample was subsequently heated, submitted to a second mercury treatment until its resistance had decreased to the new control
value $R_2 = 10^{10} \Omega$.   After appropriate cooling and forming with light,
the spectral distribution of S and σ_a was again determined.   In the
third treatment session the resistance of the sample was brought
down to $R_3 = 10^8 \Omega$.

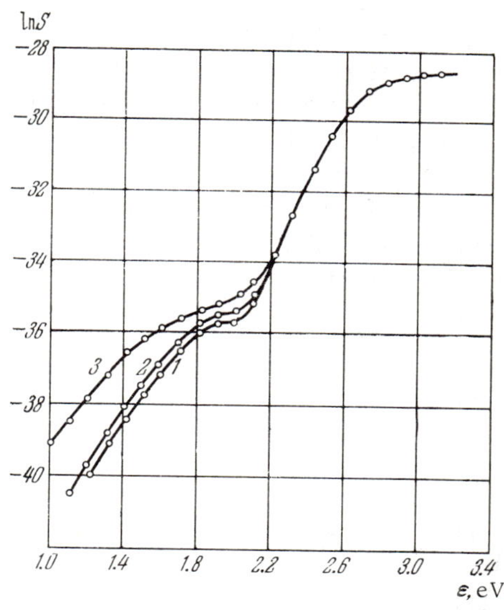

FIGURE 64. Spectral distribution of the cross-section
for various stages of activation (sample No.8VK)

The experimental values of S obtained in the three series
(sample No. 8VK) are shown on Figure 64, where the abscissa gives
the photon energy ε and the ordinate gives ln S.   It is seen on this
figure that the curves of ln S vs. ε obtained for the different degrees
of activation of the selenium film are similar to the curves of ln S
vs. ε from Chapter 3.   The only difference is that the resonance
maximum at $\varepsilon = E_r$ on these curves is less pronounced and degener-
ates into an inflexion on the ln $S = f(\varepsilon)$ curve.   All the three ln S
curves coincide in the spectral region from 2.2 to 2.9 eV, whereas
in the energy range below 2.2 eV a distinct change in the shape of
the ln $S = f(\varepsilon)$ curves is observed for different degrees of treatment.

The shape of the inflexion on the $\ln S$ curve in this spectral region changes distinctly from curve 1 to curve 3. The longer the duration of the mercury-vapor treatment, the higher and the wider this inflexion.

We recall that the different sections of the $\ln S$ curve are determined to a different extent by the s-center parameters. In the photon energy range of 2.2–3.0 eV, the dependence of S on ε is mainly determined by the parameters of the s-centers (barrier height, carrier binding energy, and width of potential well); in the energy range near the resonance level E_r, the $S(\varepsilon)$ curve depends essentially not only on the parameters of the s-centers but also on the parameters characterizing the C' levels. Since the $\ln S$ curve in the range 2.2–2.9 eV does not change as the duration of mercury-vapor treatment is increased, we conclude that the s-centers have already formed at the end of the first stage of mercury activation and that further treatment with mercury does not substantially change their parameters (although possibly increasing the number of s-centers). At the same time the C' levels of the selenium matrix change substantially with the degree of activation of the film: they become broader. To illustrate the broadening of these levels, Figure 65 plots the difference ΔS between the experimental value S and the value S_{calc} calculated from the s-center parameters derived from the $S(\varepsilon)$ curve in the range $\varepsilon \sim 2.2$–2.9 eV ($\varepsilon = h\nu$ is the photon energy).

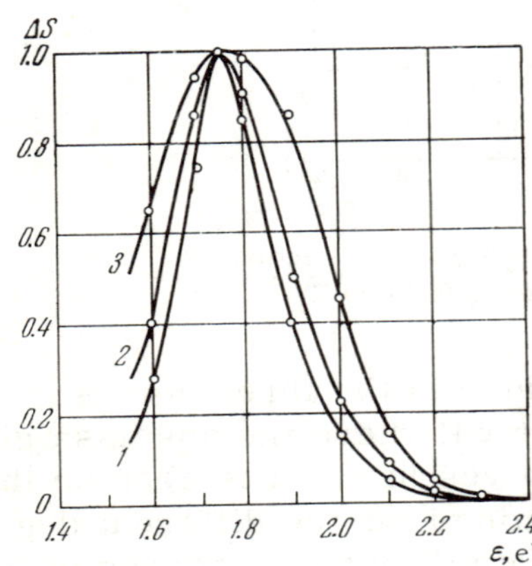

FIGURE 65. $\Delta S(\varepsilon)$ peak for various stages of activation of the selenium film (sample No.8VK)

For the sake of comparison, the scale of ΔS is given for all the three curves in relative units, with ΔS at the energy $\varepsilon = 1.77$ eV chosen as unity. As in the preceding figure, curve 1 refers to the film with $R = 10^{11}\,\Omega$, curve 2 to the film with $R = 10^{10}\,\Omega$, and curve 3 to the film with $R = 10^{8}\,\Omega$.

It is seen on Figure 65 that for all the three curves, ΔS is peaked at $\varepsilon = 1.77\,\text{eV}$. In other words, the inflexion on the $\ln S$ curves (Figure 64) represents resonant transmission of a carrier through the barrier at the level $E_{c'}$, but the resonance maximum is not as pronounced because of the broadening of the C' levels.

As the degree of activation of the sample increases, the ΔS peak becomes progressively broader. Between the first and the third stage of mercury treatment, the width is doubled. This fact confirms that the c' levels are genetically associated with mercury and that they constitute an impurity band which becomes broader as the mercury concentration in selenium is increased.

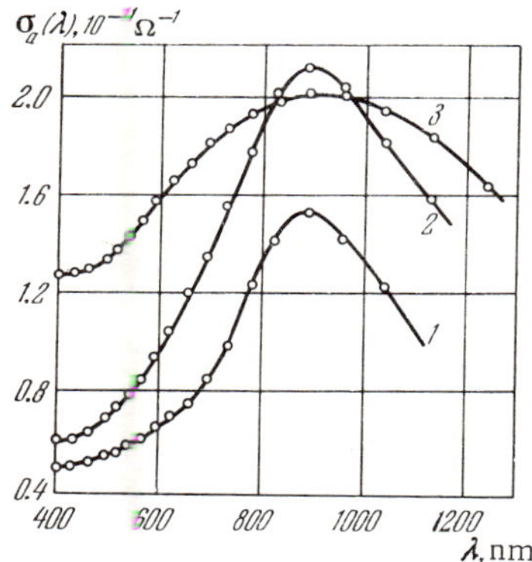

FIGURE 66. Spectral distribution of anomalous photoconductivity at various stages of activation for sample No.8VK

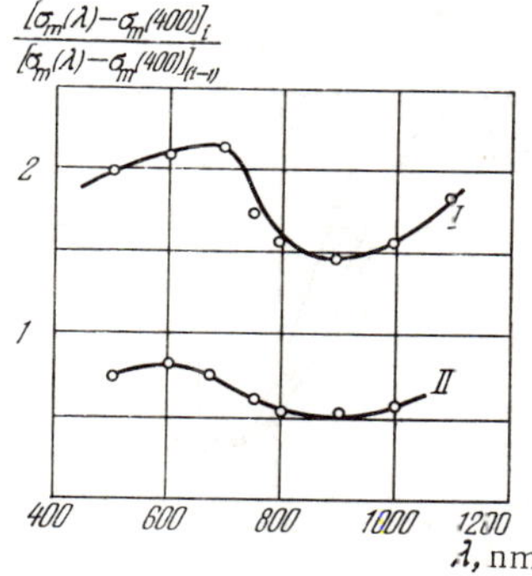

FIGURE 67. Spectral distribution of anomalous photoconductivity at various stages of activation for sample No.8VK

Alongside with the increase and the broadening of the resonance peak, the quantum yield $\beta_{o}'\,(\varepsilon)$ and the spectral distribution of anomalous photoconductivity $\sigma_{a}\,(\varepsilon)$ also change. Figure 66 shows three curves 1, 2, 3 illustrating the spectral distribution of anomalous photoconductivity after the corresponding treatment stages of the sample ($R_1 = 10^{11}\,\Omega$, $R_2 = 10^{10}\,\Omega$, $R_3 = 10^{8}\,\Omega$). On passing from the first to the second stage of mercury treatment, $\sigma_a(900)$ increases faster than $\sigma_a\,(420)$. During the third stage of mercury treatment, Z decreases both on account of the decrease in $\sigma_a\,(900)$ and especially on account of the marked increase of $\sigma_a\,(420)$. But $\sigma_a\,(\lambda)$ is made up of two parts, the conductivity due to the equilibrium carriers σ_0 and the conductivity due to the metastable carriers σ_m. In order to isolate the conductivity σ_0, we shall examine how the difference $\sigma_a(\lambda) - \sigma_a(400) = \sigma_m(\lambda) - \sigma_m(400)$, which does not contain σ_0, varies during mercury treatment. Figure 67 plots the ratio of the difference $\quad\quad$ (400) at a given stage of mercury treatment to the corresponding difference for the preceding stage of treatment as a function of the wavelength λ. Curve I represents the ratio of $\sigma_m(\lambda) - \sigma_m(400)$ after the second stage of treatment to the value of this difference after the first stage. Curve II represents the ratio of the difference after the third stage of treatment to its value after the second stage of treatment. The shape of these curves leads to the following conclusions.

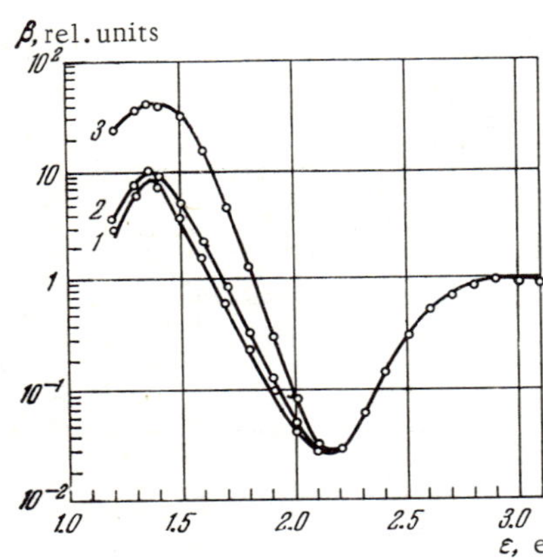

FIGURE 68. Spectral distribution of the quantum yield in sample No.8VK for various stages of activation

1. Mercury activation has a different effect on different parts of the spectrum: the short-wave region changes more markedly and the spectral region around $\lambda = 900$ nm is less affected.

2. The ratio $\sigma_m(700) / \sigma_m(400) = Z_m$ indeed varies negligibly in the process of successive mercury treatments. Thus, passing from the first to the second stage of activation $\sigma_m(400)$ increases by a factor of 2 and $\sigma_m(700)$ by a factor of 2.17. Passing from the second to the third stage, $\sigma_m(700)$ decreases by a factor of 1.4 and $\sigma_m(400)$ decreases by almost the same factor.

3. The observed changes in $Z_a = \sigma_a(700) / \sigma_a(400)$ are mainly due to the variation of σ_0, which changes with the degree of activation relatively more markedly than $\sigma_m(400)$ does.

Since $\sigma_m(\varepsilon)$ and $S(\varepsilon)$ change, the quantum yield should also change. Figure 68 shows the spectral distribution of the quantum yield β' at various stages of mercury activation of the sample curves 1, 2, 3 (for $R = 10^{11}\,\Omega$, $R_2 = 10^{10}\,\Omega$, $R_3 = 10^8\,\Omega$ respectively). The curves meet at the point $\varepsilon = 3\,\mathrm{eV}$, i.e., the value of β' for each of the curves was taken as unity at this point. The spectral distribution of the quantum yield does not change for $\varepsilon \sim 2.2-3.0\,\mathrm{eV}$, whereas in the region adjoining $1.36\,\mathrm{eV}$ the quantum yield changes quite markedly. Despite the decrease in $\sigma_a\,(900)$ between the second and the third stage, the quantum yield in this spectral region increases.

It follows from the above data that mercury has a twofold effect on a selenium film. On the one hand, s-centers are formed as a result of mercury activation. This takes place at the very beginning of the mercury treatment. On the other hand, mercury activation of selenium films produces a band of impurity levels C' with an excitation energy $\varepsilon = 1.32 \pm 0.1\,\mathrm{eV}$. As the mercury treatment progresses, the C' band widens, the cross-section for this spectral region increases, and so does the quantum yield (the cross-section increasing faster than the quantum yield). At this stage of activation, the anomalous photoconductivity begins to decrease with increasing degree of activation.

5. CONDITIONING OF SAMPLES WITH WHITE LIGHT

The treatment of the amorphous selenium layer with mercury is necessary in order to make the samples anomalously photoconductive. However, this treatment is not sufficient by itself. Mercury-treated amorphous selenium samples, even when cooled to -170°C, do not become anomalous photoconductors. Such samples must be additionally submitted to a prolonged treatment with light. To clarify this conditioning mechanism, the spectral distribution of the photo-response was investigated after mercury treatment and then after exposure to light.

As mentioned before, the spectral distribution of the photocon-ductivity of pure amorphous selenium (not submitted to mercury vapor treatment) differs substantially from the spectral distribution of anomalous photoconductivity. Figure 69 shows the conductivity of a selenium film not activated by mercury. Curve I plots the conductivity under illumination and curve II the dark conductivity. It is seen in Figure 69 that the dark conductivity does not depend on the wavelength of the previous illumination. The photoconductivity peak in nonactivated selenium is near $\lambda = 450\,\mathrm{nm}$, whereas in anomalously photoconductive samples (see Figure 6, Chapter 1), the maximum photoresponse shifts (relative to pure selenium) far into the long-wave region of the spectrum, and is situated in the infrared ($\lambda = 900-1,000\,\text{nm}$).   Consequently, the two types of spectra differ quite substantially.

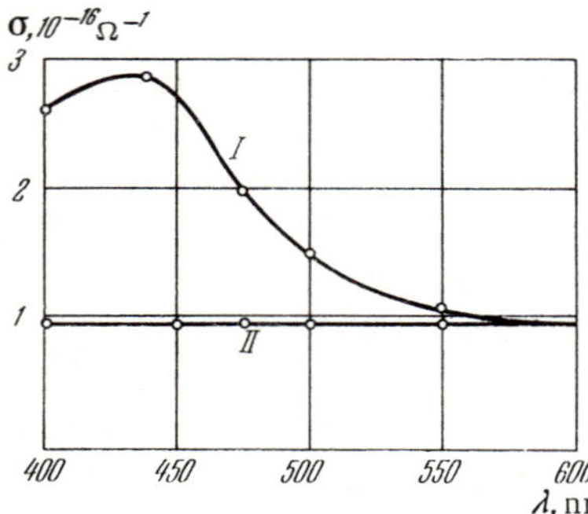

FIGURE 69. Spectral distribution of photoconductivity in amorphous selenium films not activated with mercury

Let us establish the photoresponse spectrum of amorphous selenium right after mercury treatment and cooling to quite a low temperature, before conditioning with white light, i. e., before the sample has become fully anomalously photoconductive.

The spectral distribution of the photoconductivity in a sample after mercury treatment but before conditioning with white light contains a mixture of features of both kinds of spectra mentioned above. In this mixed spectrum (Figure 70), there is a pronounced conductivity maximum at $\lambda = 450\,\text{nm}$, characteristic of pure selenium, and another maximum at $\lambda = 900-1,000\,\text{nm}$ characteristic of anomalous photoconductivity. The intensity ratio of the two maxima varies from sample to sample and depends on the degree of mercury treatment. In the given sample, the short-wave maximum of amorphous selenium prevailed.

Prolonged treatment with white light considerably changes the spectral distribution of the photoresponse. As the treatment progresses, the $\lambda = 450\,\text{nm}$ maximum decreases and after the completion of the treatment it disappears. There remains only one maximum at $\lambda = 900-1,000\,\text{nm}$, which increases with the light dose. The spectral distribution of photoconductivity established after treatment with light is presented in Figure 71 (curve I).

After the establishment of the spectral distribution of photoconductivity, this sample was heated and submitted to additional treatment with mercury and light at 130°K; the resulting spectral distribution of photoconductivity was again obtained. The result is shown in Figure 71 by curve II. The short-wave region of the spectral distribution of photoconductivity ($400-500\,\text{nm}$) did not change as a result of the additional treatment with mercury and light, whereas the photoconductivity in its long-wave region increased substantially. The photoconductivity of the sample at 600—1,000 nm increased almost by a factor of 3 (Z increased correspondingly by a factor of 3).

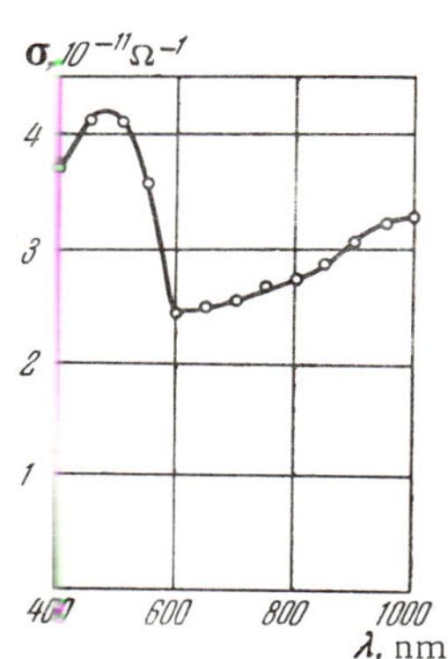

FIGURE 70. Spectral distribution of photoconductivity of mercury-treated amorphous selenium before conditioning with white light

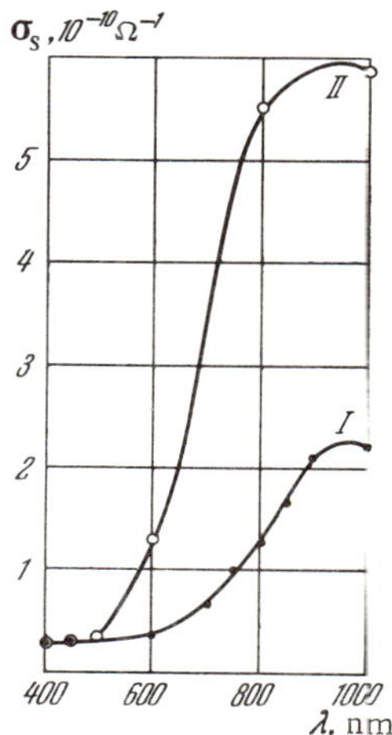

FIGURE 71. Spectral distribution of the conductivity under illumination for amorphous selenium (sample No.3VK) after treatment with mercury and white light (curve I). Curve II — after repeated treatment with mercury and light.

It follows from the above data that prolonged illumination of activated selenium films with intense white light has a twofold effect.

1. As a result of such illumination, the photoresponse of these films in the spectral region adjoining $\lambda = 900-1,000$ nm increases. The photoresponse in this spectral region is determined by the C' impurity band. Consequently, illumination promotes the formation of this band. Conditioning with light does not induce any change in the s-center parameters.

2. As a result of exposure to light, the photoconductivity of pure selenium disappears from the film. The steady-state dark conductivity of the nonanomalous current component thereby substantially decreases. It is clear that one of the mechanisms of the action of light on a selenium film consists in the redistribution of charges among the impurity centers in selenium and, as a result of charge transfer, the centers supplying free carriers in amorphous selenium are emptied. This conclusion is also in agreement with the results in thermally stimulated currents in activated amorphous selenium films.

Thermally stimulated currents (TSC) in activated amorphous selenium films were first observed by Trofimov et al. /76/. The sample was placed in a glass vacuum cryostat which was immersed in a liquid nitrogen dewar. To heat up the sample, the nitrogen dewar was removed. The current through the sample was recorded continuously.

The thermally stimulated currents were investigated for numerous samples. In all cases, the same picture emerged. There are three peaks on a TSC curve (Figure 72): peak I at $T = 130°K$, peak II at 170—190°K, and peak III at 230—250°K.

Note that the high-temperature peak III was observed upon heating a sample which had remained in the dark during the cooling phase and had not been exposed to light before heating.

The low-temperature peaks I and II appeared when the sample heated had been illuminated with a weak light flux (either blue or red) prior to heating.

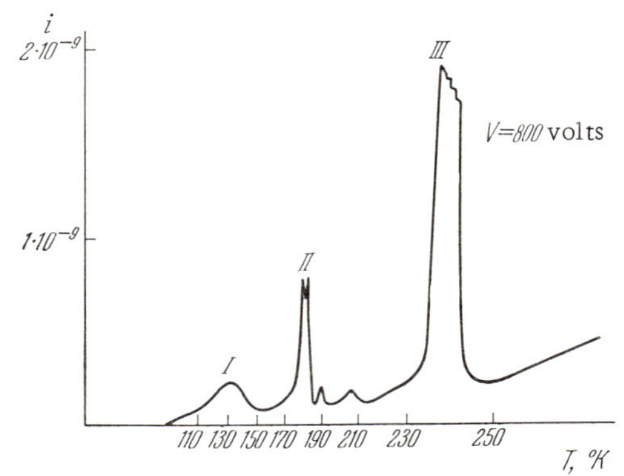

FIGURE 72. Curve of thermally stimulated current in amorphous selenium, treated with mercury

The high-temperature peak III is associated with adsorbed molecules of water or other polar materials. When the sample was placed in an acetone medium the height of this peak increased almost by an order of magnitude; when the water vapor was removed, the peak markedly decreased. The illumination of the sample with white light at a low temperature did not affect the position of this peak and its height. Consequently, this peak is not related to long-life traps.

The low-temperature TSC peaks appear only when the sample is illuminated with low-intensity red or blue light before heating.

Peak II does not always appear and not in all the samples. The energy level corresponding to the peak at 130°K was estimated at 0.47 eV from the expression in /3/. When the sample was illuminated with high-intensity white light, the height of the low-temperature peaks decreased with increasing duration of illumination, and disappeared altogether when the illumination was prolonged until the steady-state anomalous photoconductivity had been attained. This shows that these peaks are not associated with s-centers but with shallow traps in selenium, and that one of the processes taking place during conditioning with light is the emptying of the shallow trapping levels.

The process of light treatment can be controlled using the fact that exposure of a mercury-activated sample with white light leads to the disappearance of the photosensitivity maximum in the blue-violet region of the spectrum. The treatment with light should continue until all traces of photoresponse at $\lambda = 400-450\,\text{nm}$ have disappeared.

In some cases /9/, some increase in the steady-state value of anomalous conductivity was recorded when illumination with wavelength $\lambda = 400-420\,\text{nm}$ was substituted for illumination with wavelength $\lambda = 450-480\,\text{nm}$. As a rule, this points to insufficient light treatment. Further illumination of the sample with white light for a sufficient time will eliminate this rise of photoconductivity and a region of almost constant photoconductivity is obtained at 420–480 nm, characteristic of the spectral distribution of anomalous photoconductivity.

Thus, the process of forming consists of the following stages:

1. Mercury activation: s-centers are created, and the C' impurity band appears in the selenium matrix.

2. Cooling: the penetration of equilibrium carriers into s-centers ceases. The contribution of equilibrium carriers to conductivity decreases. A considerable proportion of s-centers are still filled.

3. Conditioning with light: the s-centers filled at room temperature are emptied. Charge redistribution among other impurity centers in selenium takes place, so that the characteristic conductivity of nonactivated amorphous selenium disappears.

6. CONDUCTIVITY CHANGES IN ACTIVATED AMORPHOUS SELENIUM FILMS INDUCED BY COOLING AND HEATING

Additional information on changes taking place when anomalously photoconducting films are being conditioned with light is given by the cooling and heating curves of the films, obtained before and after the conditioning process (Figure 73). Curve I shows the variation of conductivity when the sample is cooled from room temperature (point A) to 94°K (point B). Before taking the measurements of curve I, the sample was kept in the dark for 30 min at room temperature, so as to allow the s-centers to fill with carriers. The cooling was also carried out in the dark. In the dark, the carriers do not leave the s-centers, so the concentration of metastable carriers remains equal to the concentration of s-centers Q at 94°K. After cooling, the sample was heated in the dark to room temperature. It turned out that the change in conductivity follows curve I in this case also. Several cooling and heating cycles were carried out; no hysteresis of conductivity was observed if the whole cycle was conducted in the dark. It follows that there are no residual changes of carrier concentration during heating or cooling.

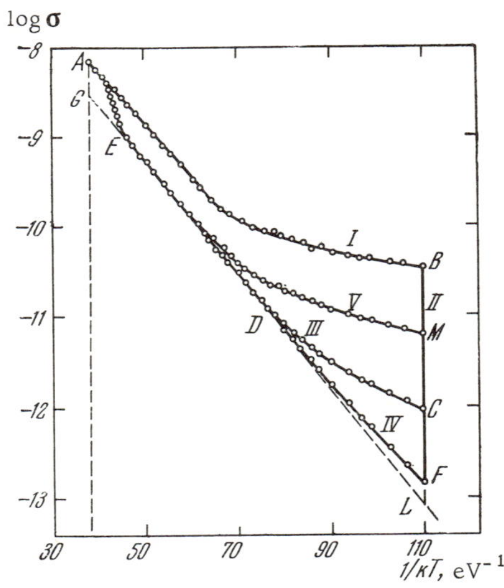

FIGURE 73. Cooling and heating curves of anomalously photoconductive selenium

However, the variation of conductivity during heating follows a different curve if the sample is conditioned with light at low temperature. This conditioning lowers the photoconductivity of the sample.

If, after conditioning, the light is switched off and the sample is kept in the dark until a steady-state conductivity is established (point C on Figure 73), and is then heated, the conductivity follows curve III.   During the cycle cooling—conditioning with light—aging in the dark—heating, the conductivity follows curves I-II-III.

If, after forming, the sample is submitted to illumination at $\lambda = 420$ nm, kept in the dark and then heated, the conductivity follows curve IV from point F (characteristic of the dark value of conductivity after preliminary illumination at $\lambda = 420$ nm) to point A. When the sample is illuminated at $\lambda = 780$ nm and then heated, the conductivity follows curve V.

The heating curves of the conductivity of activated selenium films (Figure 73) are in good agreement with the previous conclusions.

a) The dark conductivity σ_d is made up of two components, σ_0, the conductivity due to equilibrium carriers, and σ_m, the conductivity due to metastable carriers.

b) The conductivity of equilibrium carriers varies with temperature as $e^{-0.16/kT}$.

c) σ_m varies with temperature considerably less than σ_0.

d) Illumination at low temperature empties the s-centers, and the concentration of metastable carriers is reduced: the smaller λ, the larger is the reduction.

e) At $180°K$ in the dark, the s-centers slowly fill up with carriers. At $T = 250°K$, this process is fairly fast.   Below $170°K$, the concentration of metastable carriers and the degree of filling of the s-centers can be assumed to remain constant in time.

It is readily seen on Figure 73 that as σ_m decreases, the curve $\sigma\,(1/kT)$ approaches the exponential $e^{-0.16/kT}$.   The variation of conductivity on account of the filling of the s-centers by current carriers (section AE) is also quite distinct on these curves.

At low temperatures, the conductivity component σ_m predominates (it is only after illumination at $\lambda = 420$ nm that the component σ_0 becomes close to σ_m).   As the temperature of the sample approaches room temperature, the contribution of σ_0 to the dark conductivity increases substantially.   The value of σ_0 can be determined in the entire temperature interval if the section ED of the curve is extrapolated to both low (line DL) and high temperatures (line GE). Knowing σ_0, the difference $\sigma_m = \sigma_d - \sigma_0$ can be calculated.   Since the concentration of metastable carriers remains constant in the temperature range $100-120°K$, the temperature dependence of σ_m in this case is entirely due to the variation of the mobility of metastable carriers u_m.   The curve on Figure 74 shows the temperature dependence of the mobility of metastable carriers, calculated from the data of Figure 73. u_m on Figure 74 is given in relative units.   The magnitude of u_m at $T = 123°K$ is taken as unity. The points on the curve in Figure 74 correspond to the mobility u_m

calculated from the temperature dependence of the spectral distribution of anomalous photoconductivity (Sec. 8, Chapter 3). The agreement between the two series of measurements of u_m is good.

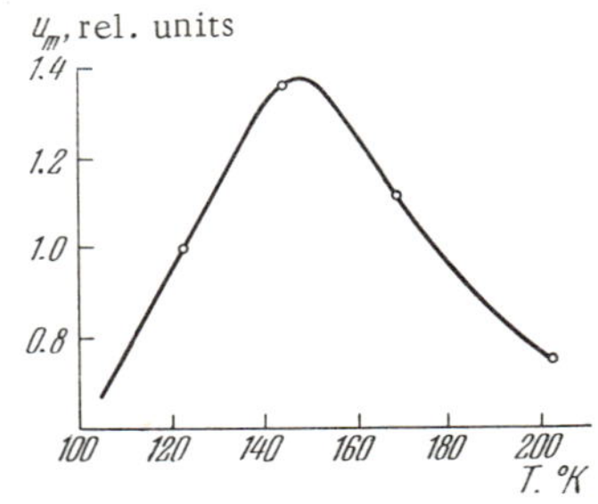

FIGURE 74. Temperature dependence
of the mobility of metastable carriers

The degree of filling of the s-centers with carriers can also be calculated from the data of Figure 73. Since the concentration of metastable carriers does not change upon cooling in the dark, it is possible to calculate eu_mQ from the value of conductivity at the point B. From the values of σ_d corresponding to points F and M (Figure 73), it is possible to determine σ_m (420) and σ_m (730), and from the ratio of these quantities to eu_mQ, we can find η (420) $= \sigma_m$ (420)$/ eu_mQ$ and η (700) $= \sigma_m$ (700)$/ eu_mQ$. For sample No. 8VK (Figure 73), illumination at $\lambda = 420$ nm emptied the s-centers almost completely; only 0.4% of them remained filled with carriers $(\eta$ (420) $= 4 \cdot 10^{-3})$. After exposure to red light $(\lambda = 730$ nm), the percentage of filled s-centers is considerably higher (16%).

7. THE EFFECT OF FILM THICKNESS ON ANOMALOUS PHOTOCONDUCTIVITY

The effect of the thickness of the selenium film on anomalous photoconductivity is investigated in /81/. The color response

$$Z = \frac{\sigma_a \ (730)}{\sigma_a \ (420)}$$

was taken as the parameter characterizing anomalous photoconductivity.

Figure 75 gives an idea of the variation of Z with film thickness l. For $l > 0.3 \mu$m, Z does not depend on the thickness. With decreasing thickness, the curve $Z = f (l)$ abruptly falls below $l = 0.1 \mu$m.

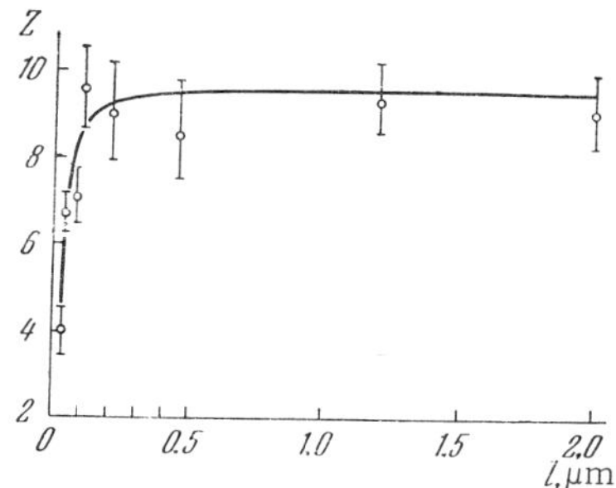

FIGURE 75. Color response of anomalously photoconductive selenium vs. the thickness of the selenium layer

The shape of the $Z(l)$ curve is explained in /81/ by the relation between three dimensional quantities: the absorption length of photons $l_\varepsilon = 1/K$, where K is the absorption coefficient of light, the diffusion length l_d, and the thickness of the sample l.

Indeed, the steady-state concentration of metastable carriers is found by setting the rate of penetration of nonequilibrium carriers into s-centers v_g (generation of free majority carriers) equal to the rate of carrier excitation by photons from s-centers v_r (a process leading to the disappearance of free carriers). Since the s-centers are situated on the surface of the selenium layer, v_r is not related to the thickness of the film; v_g is determined not only by the electrons generated directly on the surface of the layer but also by the electrons reaching the surface from a depth Δx. If the photon absorption length $l_\varepsilon > l_d$, the thickness Δx of this layer is determined by the diffusion length l_d. If $l_\varepsilon < l_d$, we have $\Delta x = l_\varepsilon$. The photon absorption length l_ε depends on the wavelength. For $\lambda = 420\,\text{nm}$, $1/K\,(420) < l_d$; for $\lambda = 730\,\text{nm}$, $1/K\,(730) > l_d$. If the film thickness $l > l_d$, the depth $\Delta x = l_d$ for $\lambda = 730\,\text{nm}$; for $\lambda = 420\,\text{nm}$, $\Delta x = 1/K\,(420)$ and Z being proportional to the product $l_d K\,(420)$ does not depend on l. If $l < l_d$, $\Delta x\,(420)$ is still equal to $1/K\,(420)$, but $\Delta x\,(730)$ is equal to the film thickness l, and not to the diffusion length l_d, and Z is proportional to $lK\,(420)$, i.e., decreases with decreasing l.

The curve on Figure 75 gave $l_d = 0.06\,\mu\text{m}$. This value of l_d is in good agreement with the diffusion length determined from the mobility u and the lifetime τ_0 of free carriers in amorphous selenium:

$$l_d = (D\tau_0)^{1/2} = \left(\frac{ukT}{e}\,\tau_0\right)^{1/2}. \tag{5.1}$$

Substituting in (5.1) $u = 5.2 \cdot 10^{-3}\,\text{cm/sec} \cdot \text{V}$ /17/ and $\tau_0 = 10^{-7}$ /12/, we find $l_d = 4 \cdot 10^{-6}\,\text{cm}$. The two values of l_d are in good agreement.

The dependence of the color response Z on the film thickness l provides another proof that anomalous photoconductivity in activated amorphous selenium films is a surface effect, restricted to a layer thickness not exceeding 0.1μm.

Let us summarize. Forming consists of the following stages.

1. Mercury activation: s-centers are created in the subsurface layer and the C' impurity band appears in the selenium matrix. In the dark, at room temperature, the s-centers are filled with carriers.

2. Cooling: the penetration of equilibrium carriers into s-centers ceases. The contribution of equilibrium carriers to conductivity decreases. All the s-centers (if the cooling is completed in the dark) or a considerable proportion thereof (if the semiconductor is cooled under illumination) remain filled.

3. Conditioning with light: the s-centers filled at room temperature are emptied. Charge redistribution takes place in impurity centers so that the typical conductivity of nonactivated amorphous selenium disappears. This charge redistribution is accomplished by short-wave light.

Chapter 6

POSSIBLE APPLICATIONS OF ANOMALOUS PHOTOCONDUCTIVITY

The specific features of anomalous photoconductivity make it possible to use anomalously photoconductive semiconductors for the construction of detectors and transducers of various physical quantities. In all these detectors, the basic property of anomalous photoconductivity is used: an anomalous photoconductor does not respond to changes of light intensity but to changes in spectral composition. Therefore, resistors using anomalously photoconductive elements do not act as photoresistors but as color-resistors (or chromatoresistors).

Some examples of possible applications of color-resistors will be discussed below.

1. RADIATION DOSE DETECTORS

As mentioned above, when the wavelength of light illuminating an anomalous photoconductor is changed, its conductivity changes. The magnitude of the conductivity increment is given by

$$\sigma - \sigma^{(0)} = [\sigma_a(\lambda) - \sigma^{(0)}](1 - e^{SLt}), \tag{6.1}$$

where $\sigma^{(0)}$ is the conductivity of the sample at the time when the illumination of wavelength λ is switched on, $\sigma_a(\lambda)$ is the steady-state anomalous photoconductivity for light of wavelength λ, S is the cross-section for carrier excitation from a trap by a photon of wavelength λ, σ is the conductivity established in the sample after illumination for a time t with light of the given wavelength, with an intensity of L photons/sec/cm^2. Since the product Lt constitutes the radiation dose, i.e., the number of photons incident in time t on $1\,\text{cm}^2$ of the sample surface, the level of conductivity of an anomalous photoconductor established as a result of its illumination is determined by the radiation dose, and not by the light intensity.

After switching off the illumination, the level of conductivity remains constant, and it is thus clear that two equal radiation doses, Lt, one of which is built up in several portions ($\Sigma \, L_i t_i = Lt$), produce the same conductivity increment $\sigma - \sigma^{(0)}$. This property of anomalous photoconductors can be employed for the construction of integrating cells used as light dosimeters /77/.

Note that the conductivity increment depends not only on the quantity of luminous energy incident on the sample but also on the spectral composition of the radiation and on the initial conductivity $\sigma^{(0)}$. In order to use an anomalous photoconductor as a light dosimeter, a number of conditions must be observed.

1. The light flux to be measured must have a constant spectral composition during the illumination phase.

2. The device will work best when the steady-state anomalous photoconductivity $\sigma_a(\lambda)$ established by the light flux is maximally different from the initial conductivity $\sigma^{(0)}$. The maximum difference in the steady-state anomalous photoconductivity is obtained at the edges of the visible spectrum, i. e., in the blue-violet and the red (more accurately, the near infrared $\lambda \approx 900\,\text{nm}$) region. The initial conductivity therefore must be established by radiation from one of these spectral regions, and the light dose be measured for the other extreme spectral region.

3. Anomalous photoconductors best perform as radiation dosimeters in the blue-violet region of the spectrum. $S(\lambda)$ has the largest value for this part of the spectrum and illumination of the sample with a light flux of this spectral composition will produce the fastest change in conductivity. The initial conductivity (the priming of the dosimeter) must be established by illuminating the sample at $\lambda \simeq 800 - 900\,\text{nm}$. If the dosimeter is intended to measure white light or light of a mixed spectral composition, with $\lambda \simeq 400 - 450\,\text{nm}$, as one of the components, the $400 - 450\,\text{nm}$ region must be isolated by means of appropriate filters, and the entire radiation dose will be estimated from the dose recorded in the spectral region $400 - 450\,\text{nm}$. The filters in this case must be properly calibrated.

To use an anomalous photoconductor as a dosimeter, it should be brought into the anomalously photoconductive state, i. e., cooled to $100 - 130°\text{K}$, conditioned with white light, and then illuminated at $800 - 900\,\text{nm}$ (if the dosimeter is intended for measurements in the blue-violet or the green region of the spectrum) or at $400 - 450\,\text{nm}$ (if the dosimeter is intended for measurements in the red or the near infrared region). The dosimeter is then ready for use.

The maximum measurable radiation dose is determined by the difference $\sigma_a(\lambda) - \sigma^{(0)}$. However, since, for values of σ close to $\sigma_a(\lambda)$, the sensitivity of the anomalous photoconductor to radiation is substantially reduced, it is advisable to restrict the dosimetry to the measurement range $\sigma \simeq [\sigma_a(\lambda)-\sigma^{(0)}]/2$. The maximum measurable dose D_0 is therefore determined by the relations

$$e^{-SD_0} = \frac{1}{2}; \quad D_0 = \frac{0.7}{S},\qquad(6.2)$$

where S is the cross-section.

After the dose D_0 has been attained, the dosimeter readings can be "erased" by illuminating it with a light flux restoring the initial conductivity $\sigma^{(0)}$, and then continue to operate until the dosimeter has been exposed to a new dose D_0. To measure a dose $D \gg D_0$, the number of switching cycles from the start to the given time is counted, multiplied by the dose D_0, and the last dosimeter reading is is added.

In dosimeters, it is desirable to use anomalous photoconductors in which the nonanomalous component is absent or is small. When a nonanomalous component is present, a considerable part of the detector readings will depend on the amplitude and recurrence frequency of the pulses.

A light dosimeter with a color-resistor can also be used to record doses of radioactive and ionizing radiation. To this end, the incident radioactive radiation should be transformed into light. The spectral composition of this transformed radiation should preferably be in the blue region of the spectrum. This transformation can be accomplished by a scintillator. There are scintillators fully suitable for this purpose.

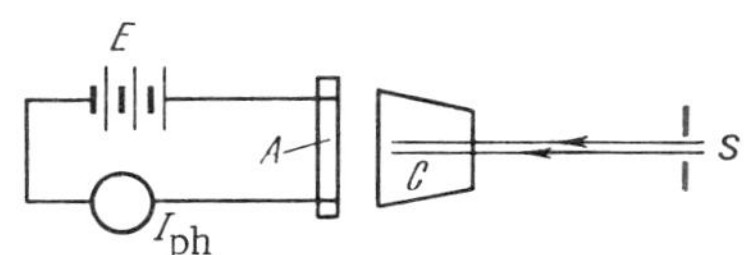

FIGURE 76. Diagram of a nuclear radiation dosimeter using an anomalous photoconductor

Directing the I_{ph} radioactive radiation on a scintillator, and the light output of the scintillator on an integrating cell with an anomalous photoconductor, we can find the radioactive radiation dose by proper calibration of the current in the cell.

Figure 76 shows a diagram of a radioactive radiation dosimeter using an anomalous photoconductor as a detector. S is the source of radioactive radiation, C the scintillator, A the anomalous photoconductor, E and I_{ph} respectively are the power supply and the photocurrent recorder.

We would like to stress some features of a color-resistor dosimeter.

1. The detector has no "Z inertia."

2. Since the relaxation time of anomalous photoconductivity is large and exceeds one and a half years, the dosimeter can record light fluxes of very small intensity accumulated during a long time.

3. Since the anomalously photoconductive properties are also preserved when anomalously photoconductive selenium is illuminated with light of very high intensity, the detector is suitable for the integration and dosimetry of isolated giant pulses.

4. The detector is suitable for dosimetry of pulses and radiation fluxes in the near infrared $(750-900\,\text{nm})$.

5. The detector is almost insensitive to the direct action of γ-radiation.

6. When a detector with a color response $Z \simeq 100$ is included in a bridge circuit, it will integrate very small doses of light. Doses of $10^{-10}\,\text{W/cm}^2$ can be recorded even if it takes a day or longer for such a dose to build up.

7. A color-resistor can accumulate information on extraneous light stimuli when no electrical supply is connected to it.

2. COLOR DETECTORS

Since an anomalous photoconductor reacts to the spectral composition of light and not to its intensity, it is possible to construct a number of devices using such semiconductors as color detectors.

a) Radiation color-temperature detector

As mentioned before (Sec. 5, Chapter 2), when an anomalous photoconductor is simultaneously irradiated with two light fluxes, the resulting level of photoconductivity depends on the ratio of the intensities of the light fluxes. It is advisable to choose one flux from the spectral range $\lambda \simeq 400-500\,\text{nm}$ and the other from the range $\lambda_2 = 700-900\,\text{nm}$. For these wavelengths the ratio

$$Z = \frac{\sigma(\lambda_2)}{\sigma(\lambda_1)}$$

is maximum. The conductivity is determined by the relation

$$\sigma = \sigma_1 \frac{1 + a\,(L_2/L_1)}{1 + b\,(L_2/L_1)}, \qquad (6.3)$$

where σ_1 is the steady-state conductivity established by illumination with one of these fluxes; $a = K_2\beta_2/K_1\beta_1$ and $b = S_2/S_1$ are parameters which depend on the wavelengths employed. a and b can be determined in advance by illuminating the color-resistor with each of the light fluxes separately. Parameters a, b and σ provide the full specifications of the device. If a and b are known, the unknown ratio $x = L_2/L_1$ can be determined from the relation

$$x = \frac{\sigma - \sigma_1}{\sigma_1 a - \sigma b}. \qquad (6.4)$$

The color temperature of a radiation source can be determined from the intensity ratio of the red and blue spectral components /78/.

TABLE 8

T, °K	$x = \dfrac{L(730)}{L(420)}$	σ/σ_1
1,160	36,000	35
2,320	84.5	2.78
3,480	11.1	1.24
5,800	1.3	1.05

Table 8 gives the relative values of the conductivity of a color-resistor σ/σ_1, which appear under illumination with two light fluxes of $\lambda_1 = 730$ nm and $\lambda_2 = 420$ nm (Figure 77) isolated by the filters Φ_1 and Φ_2 from a black body source at a temperature of T °K; the table also gives the intensity ratios

$$x = \frac{L(730)}{L(420)}$$

in the red and blue regions. The color response Z of the color-resistor, presented in Table 8, is 36.6.

It is seen from Table 8 that the peak response of the color-resistor falls in the temperature interval 1,160—2,320°K. In this temperature range, a change of 1° causes the conductivity of the color-resistor to change by more than 1%. Consequently, such changes of temperature can be recorded with certainty.

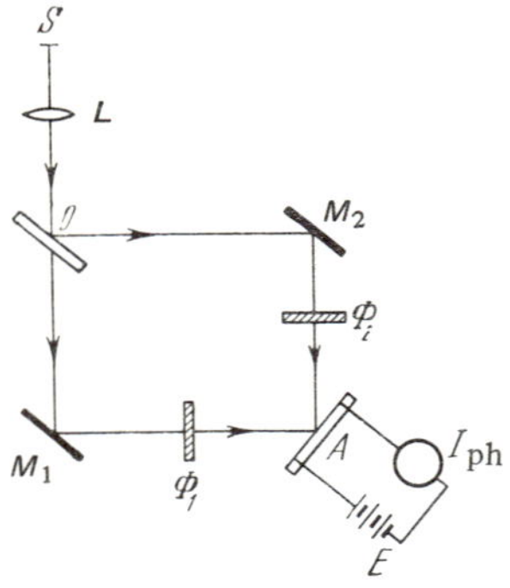

FIGURE 77. Diagram of an anomalously photoconductive color-temperature detector

At higher temperatures, $2,000-2,500°K$, the response of the color-resistors is substantially lower; however, in this case too, the response is quite high, and a 1% change of conductivity corresponds to a 5% change in the radiation source temperature.

It is possible to prepare color-resistors with a higher color response than that of the sample presented in Table 8 (see /78/).

When necessary, color-resistors can be prepared for fairly accurate measurements of the temperatures of black or grey radiation sources of the order of a hundred thousand degrees.

One of the possible designs of a color-temperature detector is shown in Figure 77. Here S represents the light source whose temperature is to be determined, L is a lens forming a parallel beam of light, O is a semitransparent mirror, M_1 and M_2 are reflecting mirrors or prisms of total internal reflection, Φ_1 and Φ_2 are filters isolating different spectral components, A is an anomalous photoconductor. E is a d. c. power source, I_{ph} is an instrument measuring the current through the detector (the scale of this instrument can be graduated directly in degrees).

The readings of a system which measures the color-temperature by means of an anomalous photoconductor do not depend on the intensity of the light flux used for the measurement, and are entirely determined by the spectral composition. A pyrometer with a color-resistor is distinguished by its high sensitivity and by easy compatibility with various automatic systems.

b) Colorimetric detector

The anomalous photoconductivity property expressed by relation (6.4) can be used to construct a detector for direct colorimetric measurements. For this purpose, the detector is exposed to two fluxes: the radiation flux of the given spectral composition whose intensity is to be measured and a secondary background light flux of a certain wavelength and intensity.   The composition of this secondary illumination can be selected by an appropriate band-pass filter so as to make the measurable range of intensity variation of the given spectral composition (given color) the widest possible. The intensity of the given spectral band will be determined by the photocurrent induced in the color-resistor by the radiation being investigated and by the secondary background illumination.   According-ing to (2.73) the photocurrent is a measure of the intensity ratio of these two fluxes and, for a fixed background illumination, is thus a measure of the intensity of the unknown radiation component.

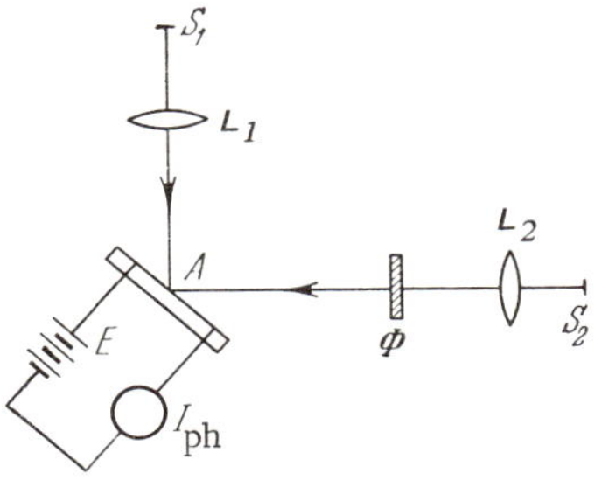

FIGURE 78. Diagram of a colori-metric detector

One of the possible designs of a colorimetric detector is shown in Figure 78. Here S_1 is the source to be subjected to colorimetry, L_1 is a lens which forms the radiation from this source into a paral-lel beam, A is the anomalously photoconductive detector, S_2 is a standard light source (the secondary background illumination), L_2 is a lens forming the background illumination into a parallel beam, E and I_{ph} are the power source and the current recorder in the measuring circuit of the detector.

c) Detector for the determination of the spectral composition of radiation

As shown above, the steady-state photocurrent in an anomalous photoconductor does not depend on the intensity of illumination but is determined by its spectral composition. Making use of this property of anomalous photoconductivity, it is possible to construct a detector for the determination of the spectral composition of radiation.

If the radiation investigated is monochromatic, the wavelength can be determined from the photocurrent through the detector. If the radiation investigated has a mixed spectral composition, the intensity ratio of its components in three spectral regions can be measured.

A light flux of complex composition can be decomposed into three light fluxes from different spectral regions, say the red, the blue and the green. The relative intensities of the component fluxes will characterize the spectral composition of the mixed radiation flux.

Thus the spectral composition of radiation can be characterized by measuring the ratio of red to blue and red to green. A diagram of a system for such intensity ratio measurements of two spectral regions has already been given above. It is clear that if one of the filters (Φ_1 say), which selects the blue spectral region, is exchanged for a green filter, two readings taken with the instrument (one with blue filter Φ_2 and another with green) will give a quantitative characteristic of the radiation spectrum in the form of the ratio $L_{red} : L_{green} : L_{blue}$. It is essential that the measured value of this ratio should not change if the intensity of the light flux changes in the course of the measurements (the spectral composition remaining constant).

3. MEMORY ELEMENTS BASED ON ANOMALOUS PHOTOCONDUCTIVITY

One of the main properties of anomalous photoconductivity is the prolonged retention (memory) of the photocurrent. Since the photocurrent is determined by the spectral composition of radiation, an anomalous photoconductor possesses spectral memory, i. e., it remembers the spectral composition of the incident light. Making use of this property of anomalous photoconductivity, various detectors of spectral and optical memory can be constructed.

In order to include all the colors of the visible spectrum in the memory of such a detector, the photocurrent induced in the detector by infrared light ($\lambda \simeq 900-1,000\,nm$) can be taken as the zero level. Infrared radiation induces a very strong current in anomalously photoconducting selenium, roughly double the photocurrent induced by illumination at $\lambda = 730\,nm$. This infrared light ($\lambda = 900-1,000\,nm$) also should be used to erase the signal (light stimulus) stored in the detector memory.

An anomalously photoconductive detector can be used as a
spectral memory element when it is necessary to store the color
of an object. Such detectors can also be used as optical memory
elements. Such memory elements are controlled by light fluxes
of definite spectral composition. There may be two such fluxes of
different composition if binary storage is used and three fluxes for
ternary storage. In principle, quaternary or higher-order storage
are possible.

4. ANOMALOUS PHOTOCONDUCTIVITY AT ALTERNATING VOLTAGES

Almost all published work on anomalous photoconductivity has
been carried out with d. c. voltage in the photoresistor circuit.
Consequently, the anomalous direct current has been investigated.
However, for many detectors, a.c. voltage is needed to feed the
measuring circuit.
Since the relaxation of anomalous photoconductivity in mercury-
treated amorphous selenium layers is slow (the color-resistor has
some inertia), some new features (for example, certain characteristic times) may appear in a.c. anomalous photoconductivity. It
is therefore necessary to study the properties of an anomalous
photoconductor supplied by an a.c. voltage source.

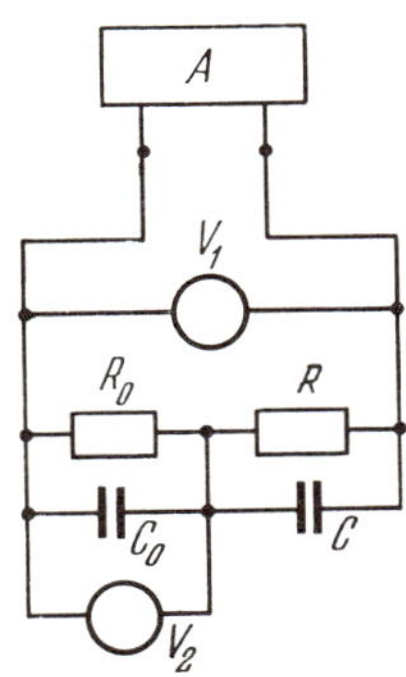

FIGURE 79. Circuit for
the investigation of anomalous photoconductivity at
a.c. voltage

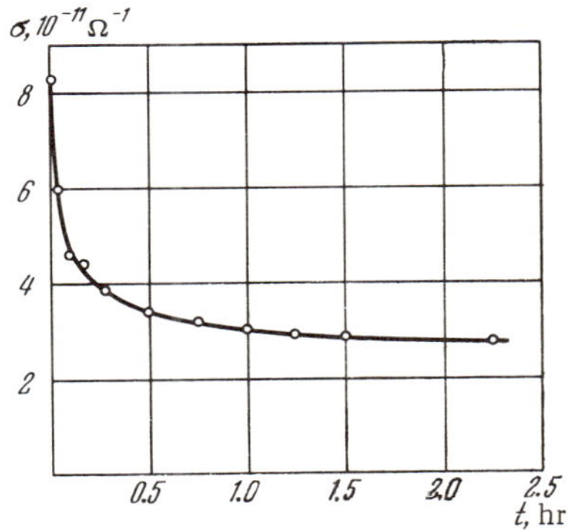

FIGURE 80. Variation of conductivity
(sample No.T1) during forming at a.c.
voltage

The investigation of anomalous photoconductivity under a.c. voltage was carried out in /79/ using the circuit shown in Figure 79. In this diagram A is the a.c. voltage source, R_0 and C_0 are standard resistor and capacitor, R and C are the resistance and capacitance of the sample. The photoconductivity was determined from the voltage drop across the standard resistor (voltmeter V_2) and the voltage applied to the circuit (voltmeter V_1). Since R_0 was taken much smaller than the resistance of the sample R, the sample conductivity was determined from the relation

$$\sigma = \frac{1}{R_0}\frac{V_2}{V_1}.$$

Samples of mercury-treated amorphous selenium exhibited distinct anomalous photoconductivity at a.c. voltage. As in the case of d.c. voltage, a.c. anomalous photoconductivity appears after preliminary treatment with light at $T < 200°K$. Figure 80 shows a curve of the conductivity variation (sample No. T1) in the process of its "forming" with white light. At 20 W lamp with a 5 cm thick Plexiglas filter served as light source. An a.c. voltage of 6−10 V, $f = 1,000$ Hz was applied to the sample. The variation of the sample conductivity during treatment with white light at a.c. voltage is generally the same as at d.c. voltage. Exposure to white light lowers the conductivity of the sample, which tends to a certain limit. When this limit is reached, the sample is in the anomalously photoconductive state.

No difference whatsoever in the forming of the sample with white light was observed at a.c. and d.c. direct voltage.

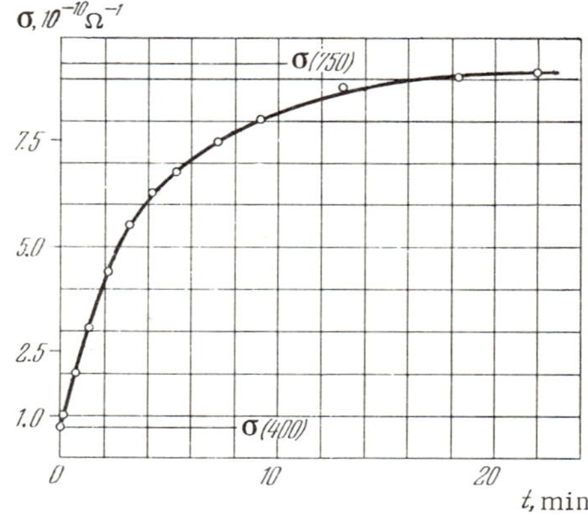

FIGURE 81. Conductivity relaxation curve for illumination at $\lambda = 750$ nm

Figure 81 shows a conductivity relaxation curve corresponding to the illumination of the sample with $\lambda = 750$ nm light fed under a.c. voltage of 10 V 1 kHz. Before the illumination, the sample was kept in the dark following previous exposure at $\lambda = 400$ nm. The induced conductivity is denoted in the figure by $\sigma(400)$. The magnitude $\sigma(750)$ represents the dark conductivity established following illumination at $\lambda = 750$ nm.

As in the case of d.c. voltage, different steady-state conductivities are obtained after illumination with red and blue light. The a.c. value is close to the corresponding d.c. value.

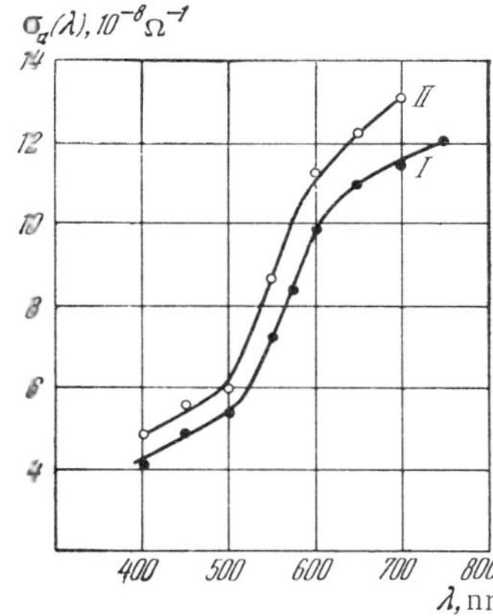

FIGURE 82. Spectral distribution of anomalous photoconductivity under d.c.(curve I) and a.c. (curve II) voltage

The spectral distribution of a.c. anomalous photoconductivity was also studied. Figure 82 shows two spectral distribution curves of anomalous photoconductivity. One of them (curve I) is the spectral distribution of anomalous photoconductivity under d.c. voltage. Curve II is the spectral distribution of anomalous photoconductivity measured under a.c. voltage of frequency $f = 1$ kHz. The curves are similar, and differ only in the somewhat higher value of $\sigma_a(\lambda)$ at a.c. voltage. This is apparently due to the shunting of the sample by the capacitive impedance under a.c. voltage, as a result of which its effective resistance is lowered.

When the a.c. frequency was increased, both $\sigma_a(\lambda)$ and the difference $\sigma_a(\lambda_1) - \sigma_a(\lambda_2)$ decreased. It was established that this decrease was purely a circuit effect, due to the capacitive impedance present in the measuring circuit. In order to identify the changes taking place in the sample itself under a.c. voltage, the influence of frequency on the active resistance of an illuminated sample was investigated. Figure 83 shows the dependence of the active

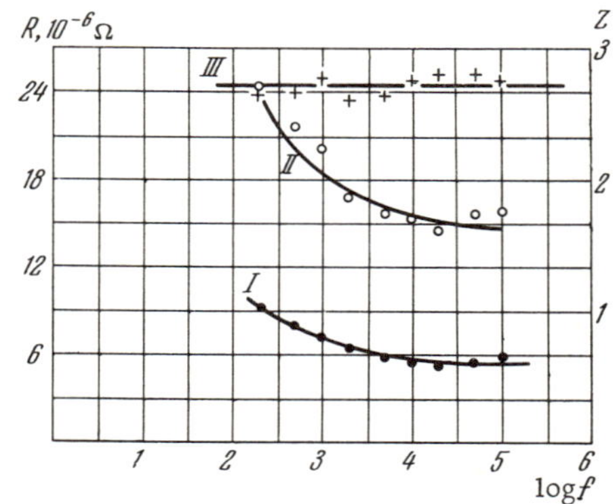

FIGURE 83. The active resistance of an anomalous
photoconductor vs. a.c. frequency

resistance of the sample R on the logarithm of frequency, when the
sample is illuminated at $\lambda_1 = 400\,\text{nm}$ (curve I) and $\lambda_2 = 740\,\text{nm}$
(curve II). As the frequency increases, R decreases, tending to a
fixed limit for $f > 10^5$ Hz. This means that, as the frequency is
increased to $f \sim 10^5$ Hz, the anomalous photoconductivity grows both
in red and blue light. Curve III on the same figure represents the
ratio $Z = \sigma_a(\lambda_2)/\sigma_a(\lambda_1)$ as a function of frequency. The ratio Z is
frequency-independent. The field frequency thus does not affect
the concentration of current carriers in long-life traps, but rather
the transport conditions of metastable carriers, i.e., the mobility u.
It follows that detectors can be designed utilizing the phenomenon
of anomalous photoconductivity, and yet operating at alternating
current.

CONCLUSION

The results presented in this book show that mercury-activated amorphous selenium films possess a wide range of unusual photo-electric properties. In these films, various slow relaxation processes occur at both low and room temperatures. At temperatures of 100—170°C, the photoconductivity induced in these samples by illumination depends only on the spectral composition (and not on the intensity of light) and is retained for an indefinitely long time after the illumination is switched off (spectral memory). When the temperature is raised in the dark to 170—200°K, the conductivity of a sample which has been previously illuminated at 120°K is observed to increase. Such a conductivity increment with increasing temperature is also characteristic of thermally-stimulated conductivity, but in activated selenium films this conductivity growth may persist for a very long time (often many tens of hours), and the conductivity increment attained is retained for an indefinitely long time.

At still higher temperatures, various complicated types of photoconductivity are observed in the films. These phenomena cannot be explained by polarization processes, since they are also observed at a.c. voltage of quite high frequencies (up to 10^5 Hz); they cannot be due to processes associated with the electrodes barriers, since they are observed for a wide variety of electrodes both with higher and lower work functions than that of selenium. They are not associated with charge injection into selenium from the electrodes, since these phenomena are clearly observed in the region where Ohm's law is valid. Anomalous photoconductivity is also observed in samples with a resistance of $\sim 10^4\,\Omega$.

It is shown in the book that all these phenomena can be explained by the existence of special impurity centers — long-life traps (storage centers or s-centers) in the films.

An s-center is a microscopic formation, a "subcolloidal" particle. The nature and the mechanism of these formations are not quite clear at this stage, and only lately have some papers appeared showing that, in a number of cases, the nucleation of such particles is possible from purely thermodynamic considerations (A. E. Glauberman, M. A. Krivoglaz et al.). Studies of the spectral distribution of photoconductivity and of the cross-section for the excitation of carriers from s-centers constitute one of the effective methods of investigating their properties, since this approach yields the potential function of the centers.

The study of the properties of s-centers is at its very beginning. It is not clear what materials, in addition to selenium, may contain such centers.   The mechanism of formation of s-centers has not been identified and it is not known what substances in addition to mercury are conducive to their formation.

It is well known that in a number of materials (copper oxide, PbSe, and others) the appearance of negative photoconductivity is accompanied by the formation of colloidal particles.   It is almost beyond doubt that these materials must contain s-centers, and it is only because the appropriate methods of investigation — i.e., measurements of the spectral distribution of the cross-section for the excitation of carriers from such centers, in a wide temperature range — have not been used that the existence of s-centers in these materials have escaped the attention of investigators.

In our opinion, urgent studies of the widest possible range of potential materials with s-centers should be launched, and we hope that the publication of this book will stimulate interest in these materials and the development of new semiconductor devices using anomalous photoconductors.

BIBLIOGRAPHY

1. Ryvkin, S.M. Photoelectric Phenomena in Semiconductors.— Fizmatgiz, 1963. (Russian)
2. Newman, R. and W. Taylor.— Solid State Phys. **8** (1959), 49.
3. Bube, R.H. Photoconductivity of Solids.— New York, Wiley, 1960.
4. Stil'bans, L.S. The Physics of Semiconductors.— "Sovetskoe Radio," 1967. (Russian translation)
5. Korsunskii, M.I., N.S. Pastushuk, and G.D. Mokhov.— Fiz. Tverd. Tela **3**, No.9 (1961), 2667.
6. Pastushuk, N.S. Electroconductivity of Amorphous Selenium.— Cand. thesis. Kharkov, Polytech. Inst., 1965.
7. Korsunskii, M.I., A.D. Volchek, K.S. Garger, and V.V. Klimenko.— Izv. AN Kazakh SSR, phys.-math. ser., No.2, 1969.
8. Korsunskii, M.I. and N.S. Pastushuk.— Fiz. Tverd. Tela 6 (1964), 254.
9. Korsunskii, M.I., O.A. Trofimov, K.S. Garger, and D.K. Daukeev. — Izv. AN Kazakh SSR, phys.-math. ser., No.2 (1966), 76.
10. Korsunskii, M.I., A.D. Volchek, K S. Garger, and V.V. Klimenko.— Doklady AN SSSR 183, No.1 (1968), 71.
11. Hartke, J.L. and P.J. Regensburger.— Phys. Rev. 139A (1965), 971.
12. Moss, T.S. Optical Properties of Semi-conductors.— New York, Academic Press, 1959.
13. Rose, A.— Phys. Rev. 97 (1955), 322.
14. Rose, A. Concepts in Photoconductivity and Allied Problems.— New York, Interscience Publishers, 1963.
15. Korsunskii, M.I.— Fiz. Tverd. Tela 3, No.10 (1961), 3181.
16. Korsunskii, M.I.— Izv. AN Kazakh SSR, phys.-math. ser., No.3 (1963), 31.
17. Spear, E.— Proc. Phys. Soc. B70 (1057), 669; B76 (1960), 826.
18. Korsunskii, M.I. and K.S. Garger.— Izv. AN Kazakh SSR, phys.-math. ser., No.2 (1965), 23.
19. Volchek, A.D., K.S. Garger, and M.I. Korsunskii.— Fiz. Tverd. Tela 8 (1966), 1625.
20. Korsunskii, M.I., E.A. Grechko, and A.I. Khrol'.— Izv. AN Kazakh SSR, phys.—math. ser., No.2 (1963), 13.
21. Korsunskii, M.I., A.D. Volchek, and K.S. Garger.— Izv. AN Kazakh SSR, phys.-math. ser., No.6 (1967), 61.
22. Nabitovich, I.Ya.— Nauchnaya i Prikladnaya Fotografiya 11, No.2 (1966), 95.
23. Volchek, A.D.— Cand. thesis. Alma-Ata, Kazakh State Univ., 1970.
24. Korsunskii, M.I., A.D. Volchek, K.S. Garger, and V.V. Klimenko.— Izv. AN Kazakh SSR, phys.-math. ser., No.4 (1969), 90.
25. Korsunskii, M.I.— Izv. AN Kazakh SSR, phys.-math. ser., No.2 (1966), 83.
26. Lanyon, A.P.D.— J. Appl. Phys. 35, No.5 (1961), 1516.
27. Korsunskii, M.I. and A.N. Osipov.— Izv. AN Kazakh SSR, phys.-math. ser., No.2 (1969), 69.
28. Lanyon, A.P.D.— Phys. Rev. 130 (1963), 134.
29. Dresner, J.— J. Chem. Phys. 35 (1961), 1528.
30. Gilleo, M.A.— J. Chem. Phys. 19 (1951), 1291.
31. Weimer, P.K. and A.Cone.— RCA Rev. 12 (1951), 314.
32. Keck, P.A.— J. Opt. Soc. Amer. 42 (1952), 221.

33. F o r t l a n d , R.A.— J. Appl. Phys. 31 (1960),1558.

34. K o r s u n s k i i , M.I., A.D.V o l c h e k, and V.V.K l i m e n k o .— Doklady AN SSSR 193, No.4 (1970),797.

35. S t u k e , J.— Phys. Stat. Sol. 6 (1964),441.

36. K o r s u n k s i i , M.I., A.D.V o l c h e k, and V.V.K l i m e n k o.— Izv. AN Kazakh SSR, phys.-math. ser., No.2 (1971),49.

37. K o r s u n s k i i , M.I., A.D.V o l c h e k, and V.V.K l i m e n k o.— Fiz. Tverd. Tela 13, No.8 (1970),2467.

38. G l a u b e r m a n , A.E. and N.A.T s a l'.— Fiz. Tverd. Tela 10 (1968),935.

39. A d a m y a n , V.M. and A.E G l a u b e r m a n.— Fiz. Tverd. Tela 11 (1969),1912.

40. G l a u b e r m a n , A.E. and V.M.A d a m y a n.— Phys. Stat. Sol. 33 (1969),K93.

41. K i r i l l o v , E.A. Fine Structure in the Absorption Spectrum of Photochemically Dyed Silver Halide.— USSR Acad. Sci., 1954. (Russian)

42. K r i v o g l a z , M.A.— Fiz. Tverd. Tela 11 (1969),2230.

43. K u l p , B.A.— J. Appl. Phys. 36 (1965),553.

44. B r o d i e , D.E. and P.C.E s t m a n.— Canad. J. Phys. 43 (1965),969.

45. L i t t o n , C.W. and D.C.R e y n o l d s.— Phys. Rev. 125 (1962),516.

46. E a s t m a n , P.C. and D.E.B r o d i e.— Proc. IEEE 53 (1965),512.

47. K o r s u n s k i i , M.I., K.S.G a r g e r, T.A.S u l e i m a n o v, and A.G.T y s h c h e n k o.— Elektronika, No.12, 1971.

48. Z h d a n , A.G. et al.— Radiotekhnika i Elektronika 12, No.3 (1967), 569.

49. Z h d a n , A.G. et al.— ZhETF Letters 8, No.8 (1968), 402.

50. D e m i d o v , K.B. and I.A.A k i m o v.— FTP 2, No.2 (1968), 210.

51. A n d r i e s h , A.M.— In: "Issledovanie po poluprovodnikam," p.5. Kishinev, 1968.

52. V i d a d i , Yu.N. and L.D.R o z e n s h t e i n.— FTP 2, No.2 (1968), 275.

53. K o l o m i e t s , B.T. et al.— FTP 2, No.10 (1968), 1539.

54. K o l o m i e t s , B.T. and V.M.L y u b i t s.— Fiz. Tverd. Tela 2 (1968), 152.

55. P a s t u s h u k , N.S., L.B.L i t v i n o v a, M.V.R e z n i k, and M.I.K o r s u n s k i i.— Trudy Kharkov Polytech. Inst., engng.-phys. ser., 14 (1958), 111.

56. K o r s u n s k i i , M.I., N.S.P a s t u s h u k, L.B.L i t v i n o v a, G.D.M o k h o v, and M.V.R e z n i k.— In: "Fotoelektricheskie i opticheskie yavleniya v poluprovodnikakh," p.220. Kiev, Ukr. Acad. Sci., 1959.

57. R e z n i k , M.V. and M.I.K o r s u n s k i i.— Izv. vuzov, Fizika, No.3 (1960), 107.

58. K o r s u n s k i i , M.I., N.S.P a s t u s h u k, and G.D.M o k h o v.— Izv. vuzov, Fizika, No.4 (1960), 167.

59. S t o c k m a n n , F.— Zs. Phys. 143 (1955), 348.

60. K r o n g a u z , A.N. and V.K.L y a p i d e v s k i i.— ZhETF 26 (1954), 115.

61. K r o n g a u z , A.N., V.K.L y a p i d e v s k i i, and Yu.S.D e e v.— ZhETF 32, No.5 (1957), 1012.

62. L a s h k a r e v , V.E. and G.A.F e d o r u s.— Izv. AN SSSR, phys. ser. 16 (1952), 81.

63. G o m b a g , L. and J.S t e i n e r.— Acta phys. et chem., No.1 (1955), 4, 9.

64. B o r z y a k , P.G.— Ukr. Fiz. Zapiski 6 (1937), 3, 288.

65. B o r i s o v , M. and S.K y n e v.— Doklady Bulgarian Acad. Sci. 7 (1954), 21.

66. M i s s e l y u k , V.G. and E.B.M a r t e n s.— Izv. AN SSSR, phys. ser. 16 (1952), 115.

67. C h i z h i k o v , D.M. and V.P.S c h a s t l i v y i. Selenium and Selenides.— "Nauka," 1964. (Russian)

68. A b d i n o v , D.Sh., S.I.M e k h t i e v a, and G.M.A l i e v.— Zhurn. Fiz. Khim. 15, No.10 (1966), 2578.

69. S e l e z n e v a , N.A.— Izv. AN Kazakh SSR, chem. ser., No.4 (1966), 31.

70. S e l e z n e v a , N.A.— Izv. AN Kazakh SSR, chem. ser., No.1 (1969), 2.

71. S a m s o n o v , G.V. (ed.). Handbook of Physicochemical Quantities.— Kiev, "Naukova Dumka," 1965. (Russian)

72. Korsunskii,M.I., K.S.Garger, Yu.F.Stepanov,and K.Kh.Nusupov.— In: "Nekotorye voprosy obshchei i prikladnoi fiziki," p.31. Alma Ata, "Nauka," 1966.

73. Borzjak,P.G., O.E.Sarbei, and R.D.Fedorowich.— Phys. Stat. Sol.8 (1965),55.

74. Hartmann,T.E.— J. Appl. Phys. 34,No.4 (1963),493.

75. Korsunksii,M.I., A.D.Volchek, K.S.Garger,and G.P.Naumova.— In: "Voprosy obshchei i prikladnoi fiziki," p.15. Alma-Ata, "Nauka," 1969.

76. Trofimov,O.A. and N.V.Sominskaya.— In: "Voprosy obshchei i prikladnoi fiziki," p.12. Alma-Ata,"Nauka," 1969.

77. Korsunskii,M.I., K.S.Garger,and A.P.Tyshchenko.— Pribory Tekhn. Eksperim.,No.4. (1971),195.

78. Svet,D.Ya. Color Pyrometers.— "Metallurgiya," 1964. (Russian)

79. Korsunskii,M.I., K.S.Garger,and A.P.Tyshchenko.— Izv. AN Kazakh SSR,phys.-math. ser.,No.2 (1970),52.

80. Korsunskii,M.I., A.D.Volchek, and V.V.Klimenko.— Izv. AN Kazakh SSR,phys.-math. ser.,No.6 (1970),45.

81. Korsunskii,M.I., N.A.Retivov,and A.P.Tyshchenko.—Izv. AN Kazakh SSR,phys.-math. ser.,No.4 (1971),77.

82. Stuke,J.— Zs. Phys. 134 (1953),194.

83. Hilsum,C.— Proc. Phys. Soc. B69 (1956),506.

84. Gobrecht,H. and A.Tausend.— Zs. Phys. 161 (1961),205.

85. Hartke,J.L.— Phys. Rev. 125 (1962),1177.

86. Korsunskii,M.I., A.P.Tyshchenko,and N.A.Retivov.— Izv. AN Kazakh SSR,No.4 (1971),64.